THE RISE OF WOMEN FARMERS AND SUSTAINABLE AGRICULTURE

The Rise of Women Farmers and Sustainable Agriculture

Carolyn E. Sachs, Mary E. Barbercheck, Kathryn J. Brasier, Nancy Ellen Kiernan, and Anna Rachel Terman

UNIVERSITY OF IOWA PRESS, IOWA CITY

UNIVERSITY OF IOWA PRESS, IOWA CITY 52242

www.uiowapress.org
Printed in the United States of America

DESIGN BY TERESA W. WINGFIELD

The University of Iowa Press is a member of Green Press Initiative and is committed to preserving natural resources.

Printed on acid-free paper

LIBRARY OF CONGRESS CATALOGING-IN-PUBLICATION DATA

Sachs, Carolyn E., 1950– author.
The rise of women farmers and sustainable agriculture / Carolyn E. Sachs, Mary E. Barbercheck, Kathryn J. Brasier, Nancy Ellen Kiernan, and Anna Rachel Terman.
pages cm
Includes bibliographical references and index.
ISBN 978-1-60938-415-9 (pbk), ISBN 978-1-60938-416-6 (ebk) 1. Women farmers. 2. Sustainable agriculture. I. Title.
HD6077.S23 2016
338.1082—dc23 2015033642

CONTENTS

PREFACE Planting a Seed

Our first step in this journey began in the living room of a woman's farmhouse in Pennsylvania, where a group of women, which included farmers, educators, scientists, agriculture service providers involved in various aspects of sustainable agriculture, and several authors of this book, discussed the needs, possibilities, and challenges of women involved in agriculture. The energy in the room that day was electric—full of power, confidence, ideas, frustrations, dreams, and visions. We shared stories, laughed together, and exchanged practical information about an endless array of topics: extending the growing season for vegetables, raising grass-fed livestock and poultry, dealing with agriculture salesmen, and changing the world. Most of all, everyone present experienced the euphoria of spending a few hours sharing our stories with other women farmers and our supporters, and we were ready for more.

At about the same time, several of the authors worked with women's agricultural networks in Vermont and Maine to plan the first national Women in Sustainable Agriculture Conference. More than twenty women, mostly farmers and several academics (and one baby), filled two maxi-vans to make the twelve-hour drive from Pennsylvania to Vermont. This was no boring road trip, but rather a spirit-filled adventure and intense time of sharing stories, giving advice, and making long-term connections. The conversations in the vans were stimulating, entertaining, and virtually nonstop. We all learned a great deal both at the conference and on tours of Vermont farms led by local women farmers. The Pennsylvania contingent of farmers decided to hold the next Women in Sustainable Agriculture Conference in Pennsylvania. As we crossed the mountains on our way home, we came upon a field where multiple flocks of snow geese majestically gathered for a meal before flying south. Coincidently, we had just heard a talk at the conference from

WISA Conference Participants: Ann Stone, Maggie and Claire Robertson, Mary Cottone, Maryann Frazier, Amber Lockawich, Susan Alexander, Carolyn Sachs, Sandy Miller, Gabbriel Frigm, Barb Kline, Emily Cooke, Lyn Garling, Linda Moist, Clair Orner. Photograph by PA-WAgN staff.

Cherokee activist Pamela Kingfisher about how leaders in her tribe rotated much like the geese: when the lead goose tires, another goose seamlessly flies into the lead, then the next, and the next. At the very beginning of our new journey together, the women farmers in the vans decided we would lead like geese to reach our desired futures.

Inspired by these two experiences, this book captures the energy, vision, struggles, and successes of women farmers engaged in sustainable farming. Over the course of ten years, the authors of this book collaborated, cooperated, and engaged with women farmers to accomplish many actions: form a vibrant network, research the challenges and opportunities for women farmers, facilitate farmer-led educational programs, meet women farmers from other regions and countries, support each other in our efforts, and learn from each other as we try to create a more environmentally sound, economically viable, and socially just food and agriculture system.

Through our work with women farmers, several of the authors of this book formed a research and outreach team to understand the barriers and opportunities of women farmers and to build an educational program based on our research findings to meet their needs. We started doing the research reported in this book for very practical reasons—to see what information, resources, and activities women farmers needed to succeed. Over time, we wove together research, networking, and educational programs for women farmers while contributing to academic conversations on gender and agriculture. As researchers, we all are currently, or were previously, faculty members and graduate students in land-grant universities in colleges of agriculture. Even though serving the needs of agriculture is at the foundation of the land-grant mission, we have been challenged by those beholden to a traditional view of farming because of our research and outreach with women in sustainable farming. We have been asked multiple times why we focus on women farmers and if they are really farmers. One would expect that in an academic setting, in an era of diversity training, institutional civil rights reviews, and general political correctness around the issue of diversity, we would have progressed beyond the need to explain women's contributions to agriculture or any profession perceived as the traditional province of men. Unfortunately, this is not the case, and even among our agricultural colleagues, women farmers are still "invisible" (Sachs 1996). Hence, we are continually reminded of the need to learn of the challenges and achievements of these women and the need to help them communicate these challenges and achievements to others so that they can be appreciated, recognized, and acted upon. Given the practical foundation of our research and our commitment to advocating for women farmers, this book reflects a feminist praxis rooted in values of equality and justice in agrifood systems.

This book tells the stories of the women farmers who continually inspire us with their creativity, commitment, and generosity of spirit. We tell their stories through their voices while, at the same time, contributing to scholarship on gender and agriculture and proposing a feminist agrifood systems theory (FAST). Our ultimate goal is to facilitate social and institutional change toward a more socially just agriculture and food system, and this book is one of our contributions toward that goal. We present the broad social and institutional contexts related to gender and agriculture in which women farmers are working, and we provide analysis about the role of women in developing sustainable agriculture.

Through ten years of collaborating with women farmers, we find that they are creative and innovative in the way they confront the barriers they face both as women and as farmers engaged in sustainable agriculture. Women farmers' success depends on creating opportunities for access to resources, knowledge, and social support through nontraditional means. While these women can accomplish much on their own, their efforts are thwarted without the support of agricultural institutions at all levels to create equal opportunities for them and to remove structural and ideological barriers. Their actions represent a decentralized social movement that is leading to diversification in the food system.

From these observations we have developed FAST. Previous analysis of women and agriculture suggests that women acted within agriculture in a way that conforms to patriarchal norms. However, our theory emphasizes the innovation of women farmers who are creating alternatives to the existing systems amid structural and ideological barriers in ways that fulfill women's goals related to health, community, and environmental quality. The purpose of this book is to explore the implications of these changes for both agriculture and for women farmers, and for women in agriculture. By focusing on women farmers in Pennsylvania and the Northeast United States, we are able to draw some general and some specific conclusions about women and agriculture. We relied on several methods of data collection, which are listed and summarized in the appendix.

Readers of this book will learn about the complexity and change in farm structure and farm households regarding women's roles in decision-making, what they produce, and how they organize their enterprises. We begin by explaining how broader changes in farming and agriculture and in gender relations beyond the farm create both opportunities and barriers for women farmers. Next, we document how women struggle to identify as farmers despite how others see them. Then, we show how women are taking leadership roles in alternative food production and agricultural systems that are economically sustainable, community-based, and environmentally sound. We point to the creativity, innovation, and leadership of women farmers as they try new livelihood strategies and attempt to transform the food system. We also provide evidence of how systematic discrimination against women in agriculture leads to poor outcomes and obscures the contribution of women farmers. In all of the chapters, women farmers tell about the innovative strategies they use to tackle barriers such as limited access to land, labor, and capital and a lack of understanding of their role as farmer by their families, communities, institutions, and agencies. We also show how some agricultural institutions reinforce structural and ideological barriers for women farmers. Finally, we document how women farmers have created new networks and organizations to confront and circumvent these barriers to champion women farmers, acknowledge the problems they encounter, and provide some solutions. We conclude by drawing these themes together to explain FAST. Readers interested in food, agriculture, or gender relations will find the narratives of these women's lives to be powerful and inspiring.

We want to acknowledge the individuals who have contributed to the development of PA-WAgN programming and research. Thank you to Amy Trauger, who was instrumental in founding PA-WAgN and conducting much of the initial research represented in this book while a graduate student and postdoctoral researcher. Thanks also to Maggie Robertson, Heidi Secord, Gay Rodgers, and Lyn Garling, founding members of PA-WAgN. Thank you to Ann Stone, Linda Moist, and Patty Neiner for contributing greatly in the organization and daily operations of PA-WAgN. Thank you to Meredith Field, Madeline Franklin, Elisabeth Garner, Jennifer Hayden, Lindsay Smith, and Kathleen Wood, all who have worked as graduate assistants and contributed to the programming and research components of PA-WAgN. Thank you to the United States Department of Agriculture, the Sustainable Agriculture Research and Education Program, Beginning Farmer and Rancher Development Program, Northeast Center for Risk Management Education, Farmers' Market Promotion Program, and the Penn State Cooperative Extension Administration for financial and institutional support. Thanks to Mike McDavid, one early believer and supporter in Penn State Extension. Thanks also to Tiana DuPont, Dennis Murphy, Sam Steele, and Doug Schaefler in Extension for their programming support. Thanks to Penn State faculty Jill Findeis, Ann Tickamyer, and Jeff Hyde for their support. Finally, thank you to the many people who work with WAgN and women farmers as facilitators, mentors, and in other capacities. Although the women farmers remain anonymous for the purposes of this book, we want to thank all those who have contributed to this research through their participation.

The women farmer participants in our studies and events will be highlighted in the following chapters of the book. Below is a farmer index describing these participants and their operations. All names are pseudonyms.

Farmer Index

Barb Hartle and her husband operate a vegetable CSA (Community Supported Agriculture) on land belonging to a not-for-profit organization.

Carol Lykens and Sasha Clay farm together in the center of a major city. They raise vegetables and market them at nearby farmers' markets and through a small CSA.

Dara and Kevin Young operate a diversified vegetable, fruit, and animal operation with the help of their grown daughters, especially Sarah. They bought the small farm with the proceeds from their off-farm jobs, but now they farm full-time, selling all of their products, as well as products from neighboring farms, through a CSA and an on-farm retail store. They also offer classes on food preservation and serve meals on-farm.

Diane and Steve Perry jointly operate an herb and mushroom farm which they bought together. They specialize in on-farm educational activities for school-age children and others.

Elaine Phillips and her husband operate a hog farm. Her husband grew up on a farm, and she joined him in farming. They converted their 250-sow confinement operation to a 60-sow grass-based operation.

Freida Gordon runs a goat dairy and produces cheese for sale at farmers' markets in major metropolitan areas on the East Coast.

Haley Selkirk runs a dairy operation on land that she bought near her family's dairy farm. She also raises diverse livestock for the farm camp for girls that she operates.

Irene Tilly operates a vegetable farm on land that she leases from a public municipality. She runs a CSA and markets her own and other farmers' vegetables. She farms full-time, and recently, her husband left his job to work on the farm.

Karen and her husband, Cliff, operate a grass-based dairy. They began dairy farming on his parents' farm but have since purchased their own land. They milk about 275 cows, and their farm is over 300 acres.

Kathleen and Will Swift raise diversified livestock and grain on land that has been farmed by Kathleen's parents for many years. Her parents still live and work on the farm as do their young children. They recently converted their farm from a conventional dairy and grain farm to pastured livestock, especially beef.

Katie Inwood operates a vegetable and livestock farm associated with a private liberal arts college. She and students at the college market their products through a CSA, at a local farmers market, and to the college cafeteria.

Leisha Wagner inherited a maple sugar operation and farm camp when her husband passed away at a young age. His parents are still working on the farm, but they are now quite elderly. She operates the maple sugar operation with the help of her sons.

Leslie Durant raises vegetables on a small farm on land that she bought with her husband, who works off-farm and is not particularly involved in the farm operation.

Linda Underwood is the owner and operator of a very successful vegetable and fruit operation specializing in value-added products. She did not grow up on a farm but entered into farming by marriage. She remains the principal operator of the multigenerational farm that she inherited when her husband, Eli, passed away. She markets her products through a large CSA, through an on-farm retail store, wholesale, to restaurants and grocery stores, and through online sales.

Liz Stedman and her husband farm on reclaimed strip-mined land which they purchased together. They raise bedding plants and vegetables.

Lori Noble operates a farm that her family inherited from her grandmother. She operates a CSA, rents out the crop land, and brings disadvantaged youth to her farm in the summer months.

Lucinda Young and Bailey Decker are friends who jointly operate a draft horse farm on land that Lucinda purchased. They raise horses and hold events on their farm, including holiday events, horse rides, and other events featuring draft horses. They also produce and sell cheese from their goat milk dairy, and they offer educational activities related to spinning and knitting of wool.

Marilyn Hart operates a grass-based livestock operation where she raises diverse livestock including pigs, chickens, turkeys, and dairy heifers. She markets her products to her loyal customers at local farmers markets. Her female partner, Ruth Oleander, also farms with her. Marilyn also has a full-time, off-farm job.

Maureen North operates a vegetable and livestock operation on her farm which she inherited from her family. Her son manages the vegetable operation. They raise chickens and beef. Maureen also produces value-added baked goods and other products which she sells at the farm and at farmers' markets.

Nan and Ken Dressler jointly operate a large vegetable operation on land that they bought together. They live in a remote rural area and market their products in high-end urban areas and through a farmer cooperative that they established.

Natalie and Larry Ingram operate a large-scale organic grain farm. They also have a large-scale organic grain milling operation in the town near their farm.

Robin and Kent Green raise beef and vegetables on their small farm. They purchased the farm together and market their products through an on-farm retail store. They also operate a small meat processing operation with their partner Hank.

Tanya Nash operates a diverse livestock farm with an ever changing menagerie of animals. She raises meat goats for seasonal ethnic markets.

Farmer Updates

These thumbnail sketches of the farmers represent their situations when we interviewed them or when they participated in the focus groups or on-farm field days. Since that time, major changes have happened for a few of the farmers. We think it is important to note that some farmers continue to be successful while others are increasingly successful, and still others have confronted major challenges thwarting their success in some ways. Several farmers expanded their operations to include other enterprises or products. On the other hand, one woman's husband developed cancer, and she had to change her farming operation. Another lost her husband in a truck accident, and three couples separated or divorced, resulting in several people not being able to stay on the farm. Another farmer and her husband cannot make ends meet on the farm and may have to sell their land.

Author Index (Not Pseudonyms)

The authors engaged in feminist production and praxis in writing this book and worked collaboratively.

Mary Barbercheck is an entomologist who conducts research and extension focused on sustainable and organic agriculture, particularly in the area of biological soil health. She owns property where she keeps three donkeys and a horse, and is committed to promoting women in the sciences and agriculture.

Kathy Brasier is a professor of rural sociology at Penn State. She studies the interactions of communities and their environment.

Nancy Ellen Kiernan, faculty emeritus, evaluated a wide variety of agricultural programs as reflected in her publications in *Evaluation Review*, *Evaluation and Program Planning*, and *Journal of Extension*. Her fieldwork includes conducting focus groups among diverse groups, including the Pennsylvania Amish, landscape architects, beef and swine producers, veterinarians, and organic farmers. She codesigned and implemented civil rights reviews which focused on gender and racial participation in agricultural programs, as well as a Climate for Women study, for which she received the Achieving Woman Award from Penn State's Commission for Women.

Carolyn Sachs is a professor of rural sociology and women's studies who conducts research on women and agriculture, gender and climate change, and gender and food.

Anna Rachel Terman is a rural sociologist at Ohio University and studies the intersection of gender, race, class, and sexuality issues in rural communities, specifically in the Appalachian region of the US.

THE RISE OF WOMEN FARMERS AND SUSTAINABLE AGRICULTURE

Chapter 1

A New Crop: Women Farmers in a Changing Agriculture

The face of farming in the US is changing, and it is with increasing frequency that the farmer's face belongs to a woman. Across the nation, women are breaking new ground in agriculture by responding to renewed interests in the agrifood system with innovations in farming and by marketing farm products. The number of women farmers has increased at a phenomenal rate in recent years, and they now comprise 30% of farm operators in the US, a 19% increase from 2002 to 2007 (USDA 2009b).[1] The Census of Agriculture distinguishes between principal farm operator and other farm operators, and it is in the category of principal farm operator where growth has been even greater. Between 2002 and 2012, the number of women principal operators increased by 29% with the major increase occurring between 2002 and 2007. This number has remained relatively steady since 2007, with a decrease of all women operators by 1.6%, and women principal operators dropping by 4% between 2007 and 2012 (USDA 2014c).[2]

Even with this slight decline between 2007 and 2012, the percentage of women farm operators and principal operators has remained steady at 30% and 14%, respectively, of the total number of farm operators. The overall increase in women farmers between 2002 and 2012 likely represents a dual change for women on farms. First, the largest increase was among women between fifty-five and seventy-five years old, which could mean that women are choosing farming as a second career by purchasing, renting, or inheriting farms and are now assuming the role of principal operator. This increase may also represent a growth in the number of women on family-operated farms who have claimed the role of farmer when historically they were seldom recognized as such, deferring to their husbands, fathers, or sons as "the farmers" in the family or household. It's not clear why fewer women entered agriculture between 2007 and 2012 than between 2002 and 2007. However,

FEMINIST AGRIFOOD SYSTEMS THEORY (FAST)—KEY THEMES

Women farmers

1. create gender equality on farms amid broad societal changes in gender roles;
2. assert the identity of farmer;
3. access the resources they need to farm by pursuing innovative ways to access land, labor, and capital;
4. shape new food and farming systems by integrating economic, environmental, and social values;
5. negotiate their roles in agricultural organizations and institutions; and
6. form new networking organizations for women farmers.

because women farmers on average are older than their male counterparts, a greater percentage of women than men may have retired from farming between 2007 and 2012 and been affected by the recession.

It is in the context of this recent increase of women farmers in an occupation long dominated in the US by men that we begin our book, which is based largely on our research and work with women farmers in Pennsylvania and the surrounding states in the Northeast. In this chapter, we explore how changes in the agrifood system and changes in opportunities for women impact how women farm and how they confront the challenges and barriers to farming. By the final chapter, we draw on this work and the stories of these women farmers to craft a feminist agrifood systems theory (FAST). This theory provides a framework to help explain why women face barriers in farming, how they respond to these barriers, and how they are creating new models of farming. FAST simultaneously examines how changes in agriculture and changes in gender relations have shifted over time in ways that both solidify and subvert traditional gender relations in agriculture. These women no longer fully embrace the traditional primary role of the farm wife within a heterosexual marriage, but rather are assuming a broader range of roles in agriculture and providing both a critique and an alternative to the conventional and patriarchal agricultural system (see box 1).

Here, we lay the groundwork by describing trends both within and outside of agriculture as a context in which women-operated farms are emerging and reshaping the agrifood system. We start by describing the general barriers for women that stem from patriarchal systems and argue that recent changes in agriculture and in gender relations in society as a whole have opened opportunities for women to negotiate these barriers. As a result, women have responded to new opportunities in farming with creative and innovative adaptations to surmount the barriers that still exist. These adaptations are creating space for alternative agricultural and food systems. This assertion sets the context for this book.

In this chapter, we describe four sets of issues within the traditional agricultural system in the US that create significant barriers for women's full participation in agriculture and their claiming identity as "farmer." These include (1) the persistent legacy of patriarchy on family farms, (2) substantial financial barriers for entry into conventional commercial agriculture, (3) increased use of capital-intensive technologies on commercial farms, and (4) the enduring sexism in agricultural institutions. Concomitantly, the US has seen significant growth in sustainable and organic agriculture[3] and an increasing interest in local food systems, a shift that has provided significant opportunities for women farmers in particular. We next examine trends in gender relations in the US in general, trends that highlight women's increased participation in the labor force and their growing presence as entrepreneurs in the private and nonprofit sectors. Although there have been shifts toward greater gender equality accompanying these changes, women continue to maintain responsibility for reproductive work, particularly caring for and feeding their families. Relatedly, compared to men, women tend to be more aware of, and concerned about, environmental quality, which leads to greater attention to issues of sustainability. The convergence of these trends leads to opportunities for women to creatively engage in agriculture in ways previously not open to them and to participate in reshaping the US agrifood system in ways consistent with their stated values related to caring, health, and their environment.

Sedimentations and Shifts in Agriculture

Persistent Legacy of Patriarchy on Family Farms Has Limited Women's Roles on Farms

Understanding why women farm and how they are involved in farming begins with acknowledging the barriers that women have experienced and continue to face in agriculture. With some exceptions, many agricultural institutions, including family farms, corporate farms, land-grant universities with their associated cooperative extension services, and agricultural agencies such as USDA, have remained resistant to addressing gender inequities. Moreover, we argue that many of these agricultural institutions remain staunchly entrenched in patriarchal ideologies, bureaucracies, and practices that reproduce and maintain barriers for women to enter and to be heard in many spheres of agriculture. Agricultural institutions continue to be male-dominated, and most of the men (and many of the women) involved adhere to patriarchal systems within agriculture.

On most traditional family farms, men identify themselves as the farmer, and women tend to be identified and identify themselves as just farm wives or helpers. Men's claim to the title of farmer is not only associated with the work they perform, but also their ownership of farmland and capital. Inheritance systems, which pass land and farms from father to son, form the core of the patriarchal system. Despite broader changes in gender equity, in general "it is farmers' sons, not farmers' daughters, who become farmers and take over ownership and management of the family farm" (Alsgaard 2013, 347). Wealth transfer to children upon the death or retirement of parents occurs through liquidation or sale of the farm, single-heir inheritance, or multiheir inheritance. Historically, farm families typically relied on single-heir systems with one son taking over ownership and operation of the farm. Farm families are moving toward multiheir inheritance systems, but dividing up assets equally is difficult when the primary assets are land and structures. Despite the move toward multiheir inheritance systems, daughters rarely take over the management of farms (Alsgaard 2013). So for many farm women, the primary path into farming is through marriage to a farmer. Of the women who have entered conventional farming, many have done so through marriage, becoming a farm wife and often remaining subordinate to men on family farms where their husbands or in-laws own the land and make decisions. These women gain financial security, status, and respectability from their position as farm wives (Fink 1992; Sachs 1996).

Many studies have provided critical accounts of the patriarchal character of the family farm. Others, however, have noted a recent movement to less traditional gender dynamics on farms (Brandth 2002; Haugen 1998; O'Hara 1998). Women increasingly reject the position of farm wife in both the US and Europe. Brandth (2002) refers to this as the "discourse of detraditionalization and diversity" (2002, 194) that runs counter to the discourse of the traditional patriarchal family farm. Rather than adhering strictly to agrarian and family farm ideologies, detraditionalization emphasizes the multiple, dynamic, and complex positions that women can have on farms.

One factor that leads to the detraditionalization of family farms is the heavy reliance on off-farm employment as a source of household income. In the US, more than half of farm operators (52%) and almost half of their spouses (45%) worked off the farm in 2004 (USDA 2012a).[4] When men work off the farm, women often take over day-to-day responsibility of running their farms—a phenomenon referred to as the feminization of agriculture (Lastarria-Cornhiel 2006; Deere 2005). In other cases, women work off the farm and may eschew their identities as farm wives as they identify more strongly with their off-farm work or occupation than with farm-related work. In addition to income through wages or salaries, off-farm work provides health insurance and other benefits critical to supporting the household.

More recent studies of gender on family farms have moved beyond studying only women's roles and include understanding men and masculinity. In a study of Kentucky farmers, Ferrell (2012) reveals that men also suffer from rigid gendered expectations and find it difficult to change their style of farming. For example, on burley tobacco farms in Kentucky, changes in labor and technology in production have resulted in tobacco farming becoming increasingly male with wives reporting little or no involvement in tobacco production. Ferrell (2012) interprets men's involvement with tobacco farming as the performance of a locally valued masculinity. At a time when government efforts encourage farmers to diversify their production from tobacco to other crops, Ferrell finds that because tobacco production is very much tied to men's masculinity and identity, switching or diversifying to other crops is a threat to their social and individual identity. In their study of Iowa farmers, Peter et al. (2000) distinguish between two types of masculinity, monologic and dialogic, which are enacted on farms. They find that larger scale commodity crop farmers adhere to a monologic masculinity that has rigid gender expectations and performances that clearly distinguish between men's and women's activities and roles on the farm. This monologic farming masculinity also approaches farming as a domination of nature. By contrast, they argue that male farmers who are engaged in more sustainable agriculture practice a dialogic masculinity with less of a need for control over nature and a greater social openness. The gender role dynamics discussed here provide a context for understanding women farmers' positions within their farm, their farming community, and the broader agricultural system. Although some changes have and continue to occur, the dominance of men and masculinity on farms is a key feature of agriculture in the US.

Financial Requirements Create Barriers for Entry into Conventional Agriculture

Currently in the US, commercial agriculture is economically dominated by an emphasis on large-scale commodity production, mostly for national and international markets.[5] The basis of this system developed after World War II with the increased commercialization, production, availability, and affordability of agricultural technologies. These technologies include synthetic fertilizers, pesticides, antibiotics, hybrid seeds, and machinery, which led to increased yields and efficiency on US farms. The adoption of these technologies enabled greater economies of scale on farms, which has been associated with farm consolidation and a trend for increasing the average size of farms over time (MacDonald, Korb, and Hoppe 2013). The shift of acreage to larger farms is part of a complex set of structural changes in commodity crop agriculture. Although most cropland was on farms with less than 600 crop acres in the early 1980s, today most cropland is on farms with at least 1,100 acres, and many farms are five and ten times that size. In 2012, the average farm size was 434 acres. This was a 3.8% increase over 2007, when the average farm was 418 acres. Middle-sized farms declined in number between 2007 and 2012. The number of large (1,000 plus acres) and very small (1 to 9 acres) farms did not change significantly during that period.[6]

Currently, the barriers of entry to large-scale, capital-intensive operations remain steep. Large-scale farms in the US produce the major commodities for the global agrifood system, the top five in 2012 being cattle and calves, poultry and eggs, corn, soybeans, and milk. Together, these commodities accounted for $261 billion in sales, or 66% of total agriculture sales. Large-scale commodity production requires considerable capital investment for equipment, buildings, and high inputs of fertilizer, chemicals, hybrid seeds, and fuel. Crops also necessitate a considerable investment in land. With commodity crop production increasingly occurring on farms with at least 1,000 acres, entry costs for a farmer are high, constituting a significant barrier for those who wish to begin farming (USDA 2014a).[7]

Animal production is also increasingly concentrated on large-scale farms. These trends affected women in particular, as the traditional products they grew and sold commercially (milk, butter, eggs, and poultry) also transitioned toward large-scale, commercial (and often male-dominated) production systems (Adams 1994). In 2012, the USDA reported a total of 60,000 dairy cow operations in the US, and of these dairy farms, 1,750 operations had over 1,000 cows accounting for 46% of the total number of dairy cows and 50% of milk production.[8] Hog production is even more concentrated.[9] Of the 69,100 hog and pig operations, 8,800 had over 2,000 hogs, accounting for 88% of hog production in 2011 (USDA 2012d). Chicken production occurs almost exclusively through contract production, in which farmers raise the birds from chick to slaughter in his and her facilities, but the birds are actually owned by the contracting corporation ("integrators"). The median number of birds produced on contract farms is 402,500 per year, but because there are a few very large operations, the typical broiler is produced on an operation of 605,000 birds (MacDonald 2008). Entry costs for a farmer in broiler production are high, with a typical house for broilers costing approximately $300,000, and many producers have multiple houses.

Access to land, capital, and technology is essential for any type of farming. Many farmers, both conventional and sustainable, obtain access to land and capital through family inheritance. Family farms typically pass from generation to generation, often with the oldest or most interested son inheriting and operating the farm. As agricultural operations have increased in size and specialization, inheritance of land and the farm from family members continues to be a major avenue of entry and access into farming. Barriers of entry for anyone are exceptionally high on these large-scale operations. For women, the barriers of obtaining land and capital are particularly high as they are less likely to inherit the farm or have to enter into difficult, unconventional negotiations with family members for land access.

We suggest these barriers contribute to the observed differences between women's and men's farms in the United States. The majority of women farmers tend to produce on small, diversified operations rather than large-scale, commodity farms. Since 1982, the majority of women-operated farms have had annual sales of less than \$10,000, and the share of women-operated farms in that sales category remains about 20 percentage points more than the share of men-operated farms with sales that low. Male-operated farms are far more likely than women-operated farms to produce the major US commodities, including grains and oilseeds and beef cattle, and somewhat more likely than women-operated farms to produce tobacco, cotton, dairy, and hogs (fig. 1).[10] In contrast, women-operated farms are far more likely than men-operated farms to have "other animals" (including horses) and "other" crops (including hay), and somewhat more likely than men to have vegetables, fruits, nuts, horticulture, poultry, and sheep and goats (USDA 2009a).

In addition to different types of products, women farmers also tend to operate smaller farms with more limited resources.[11] On average, women's farms are less than half the size of men's farms, 210 acres compared to 452 acres (USDA 2012e).[12] The discrepancy in sales is even greater, with the average sales on women-operated farms being about one-fourth of men's, \$36,440 compared to \$150,671. Part of the explanation lies in the fact that women farmers remain underrepresented in large-scale commodity production but

Type of Farm by Gender of Operator

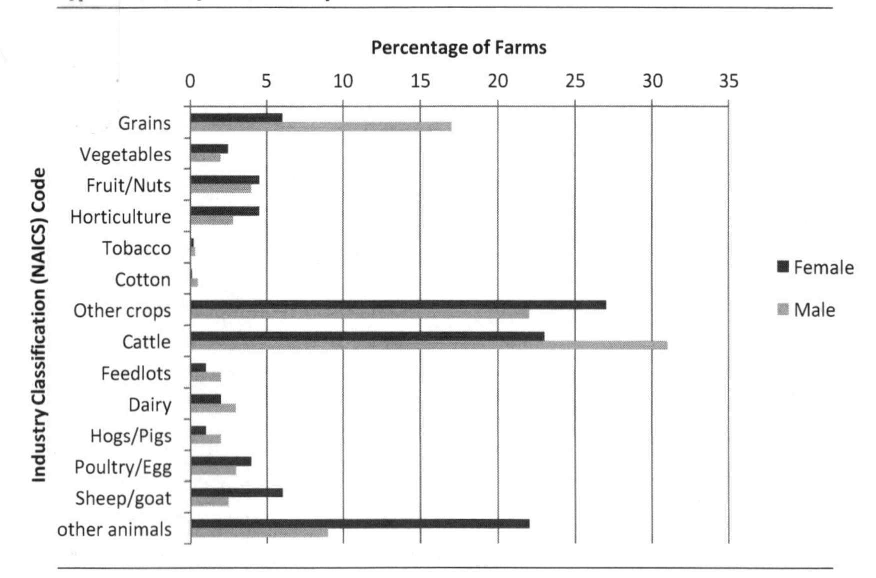

Figure 1. Types of Farm by Gender of Operator, 2007. (USDA 2009b)

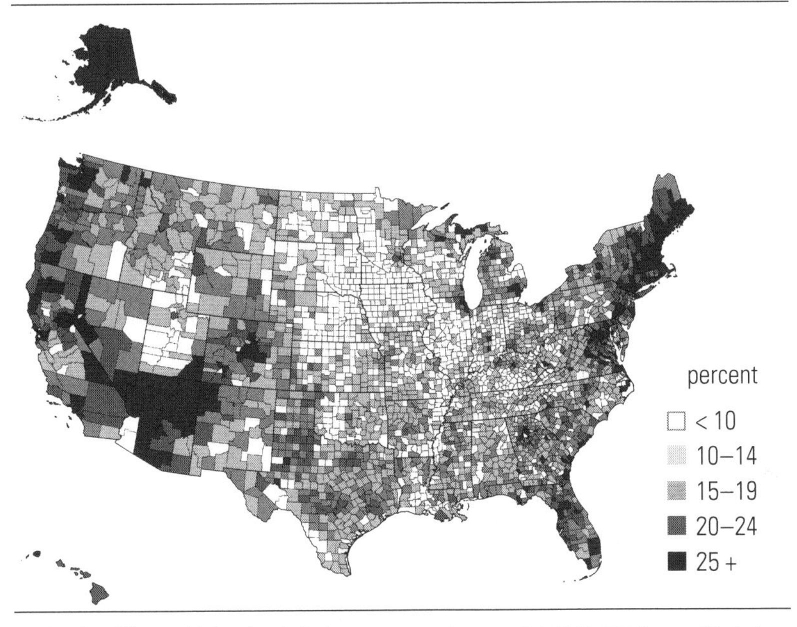

Distribution of farms with female principal operators. Source: USDA NASS, 2012 Census of Agriculture.

quite prominent in smaller scale sustainable and organic food production, selling at local markets where profit is low due to the lack of economy of scale. Women are more likely to be operating farms closer to the coasts and cities, away from the Midwest, the plains, and the South, which are dominated by large-scale crop production. However, another reason relates to the way agriculture and agricultural technologies are often gendered. In the next section we describe how institutionalized masculine and feminine identities have influenced the use of farm technology by men and women.

Technological Shifts Limit Women's Choices in Conventional Farming

Agricultural technology and masculinity are strongly linked. Feminist scholars point out that in agriculture, the tractor stands as the symbol of male identity and farm masculinity (Brandth 1995). "Farm machinery separates men from women and connects them to other men." In her analysis of tractor advertisements, Brandth found that men farmers are expected to have similar characteristics to the machine—strength, persistence, technical ability and power, and control over nature (1995, 132). Despite changes on farms in terms of gender roles, farm machinery still demarcates men and women's work. As Brandth argues, "Certain technologies are designed by and for men, presented with a masculine language and symbolism, to mark a male space, and to mediate a materialized power relation to both nature and women" (1995, 133). For many women, lack of training in using and repairing machinery presents a barrier for others to see them as farmers as well as for them to see themselves as farmers. It also inhibits their actual work on the farm and creates dependencies on others (usually men) with those technical skills.

While the tractor looms as the major symbol of technology on the farm, it is just one of the many capital, chemical, and information technologies that have transformed agriculture. Efforts to industrialize and modernize agriculture occurred during the twentieth century as farms adopted petroleum-powered farm equipment, synthetic fertilizers and pesticides, and systems for intensive animal operations. Jellison's (1993) fascinating history of women and technology on farms in the early twentieth century tracks how the emphasis on labor-replacing technologies reduced drudgery for farm women but maintained patriarchy. During that time period, farm women's labor was physically arduous, and women complained that their work was undermechanized in contrast to men's use of steam- and gasoline-powered equipment such as tractors. USDA government policies recognized the plight of farm women and were concerned that women's dissatisfaction on farms would undermine the viability of family farms as women encouraged their children to move to cities and leave farming behind. According to Jellison (1993), USDA, largely through the Cooperative Extension Service, sought to solve the problem of women's overwork on the farm through encouraging the use of modern technologies such as washing machines, electricity, and automobiles. The goal was to make farm women more like urban middle-class women without confronting patriarchal relations on the farm. However, farm women's lives differed from urban middle-class women's lives: both urban and farm women maintained responsibility for washing, cooking, cleaning, and child care, but farm women also raised chickens and other livestock, tended vegetable gardens, fed farm hands, and worked in fields and barns when their labor was needed. The new household appliances did little to change the gendered division of labor or reduce gender inequality on conventional farms. There is also evidence that "time-saving" technologies actually ended up creating more work for women because expectations for cleanliness became greater (e.g., washing clothes every time you wore them rather than after they were really dirty) (Schwartz Cowan 1983).

More recently, other types of technologies on farms, such as conservation tillage and crop hybrids genetically modified for herbicide tolerance and insect pest resistance, continue the trajectory of reducing labor demands on larger scale conventional farms. No-till farming, widely adopted by corn and soybean farmers in the US, requires less fieldwork in plowing and cultivation than traditional tillage methods through the use of pre- and post-plant herbicide application. Some herald no-till farming as a sustainable practice because it reduces soil erosion and fossil fuel use. However, this production practice also relies heavily on herbicide applications, pesticide treatments, and seeds genetically modified for herbicide tolerance and insect resistance. Clearly, this technology package reduces the need for labor (largely male labor) per acre—often leading farmers to rent or purchase more land to increase their production. Other types of technology such as large round bales for hay and wrapped round-bale silage production have also reduced labor demands on the farm. Rather than all family members, friends, and farm laborers working to get the hay in the barn before rain, one farmer (usually male) with the appropriate equipment can now handle the hay harvest. Overall, new mechanization and chemical technologies consistently lessen the physical labor requirements per area on larger scale conventional farms, leaving farmers to plant even more acreage in less time. These trends toward greater efficiency, consolidation, and mechanization together create very significant barriers to those who wish to enter farming, particularly if they do not have a family background in farming from which they can gain access to the needed land and capital.

Women who choose to enter agriculture often avoid highly mechanized agriculture or synthetic agricultural chemicals, even though small-scale sustainable farms can require more physical labor than large commodity farms. Trauger (2004) argues that women's involvement in sustainable agriculture does not actually challenge the gendered divisions of labor on conventional farms. Contrary to essentialist claims that women are choosing sustainable agriculture because it is nurturing and a kinder and gentler form of

June Hertzler of MooEcho Farms visiting dairy goats at Wayside Acres Goat Dairy during a PA-WAgN field day. Photograph by Ann Stone, PA-WAgN.

agriculture, Trauger suggests women farmers report they are doing this type of farming because "they can do the work" without the large cost of purchasing or leasing large parcels of land and machinery. The increased presence of organic, sustainable, and local agrifood systems provides an opening for women to enter farming because of their ability to enter agriculture at a lower initial cost than more highly capitalized industrial commodity production.

Enduring Sexism in Agricultural Institutions Perpetuates a Limited Role for Women on Farms

Beyond the farm, a network of public institutions (coordinated by colleges of agriculture in land-grant universities in every state) supports agriculture by providing research, education, and technical assistance. These institutions have been slow to acknowledge shifts in gender roles on farms and to adjust their programs, resulting in a lack of recognition of women as farmers and the development of assistance that would meet their specific needs.

The passage of the Smith-Lever Act in 1914 established the Agriculture and Home Economics Extension Service with the purpose of providing practical and scientific education to farmers and homemakers. At its inception, the Extension Service divided its programs into two separate spheres—the farm and the home. Extension promoted these separate spheres with men as the target audience for the farm programs and farm women as the audience for the home programs. While the USDA clearly recognized the importance of farm women to the viability of farms, they failed to recognize that farm women's work extended well beyond the confines of the home to their involvement in agricultural production (Jellison 1993).

Cooperative extension partnerships between the USDA and land-grant universities in every state continue to exist, but their emphases have shifted over time.[13] They continue to divide their educational programs into agriculture and family and consumer sciences, but they have expanded their programming into youth development and 4-H, natural resources, community and economic development, and leadership development. Despite some efforts, some level of separation of programs for men and women continues with a lack of adequate programs addressing the needs of women farmers.

The initial gendered divisions in Cooperative Extension between agricultural production (crop production) and home economics extended to research and agricultural scientists, with male scientists engaged in agronomy and other production sciences and with women holding the professorial ranks in home economics. Over time, land-grant institutions reduced the gender-segregated emphasis of their teaching as female scientists have slowly joined the ranks of male scientists in agricultural colleges. Home economics as a profession has largely been eliminated or transformed into fields such as nutrition, food science, and family studies. Men have long dominated the agricultural sciences, but women are gaining ground (Buttel and Goldberger 2002). The percentage of women earning doctorates in agricultural sciences increased substantially from 1% in 1966 to 36% in 2005 (Goldberger and Crowe 2010, 336). However, the percentage of women earning doctorates in agricultural fields continues to lag behind the biological sciences, social sciences, and humanities. In addition, women in colleges of agriculture remain drastically underrepresented in traditionally male disciplines and overrepresented in traditionally women's fields. For example, women represent only 11.5% of faculty in agricultural engineering compared to 58.8% of the faculty in nutrition (Goldberger and Crowe 2010, 344). Gender inequities remain as women are less likely to be full professors, are more likely to be untenured, receive significantly less pay, and have fewer ties to industry. Thus, research, teaching, and knowledge production in agricultural science generally tends to remain male-dominated.

Another institution that heavily impacts agriculture and agricultural policy is the United States Department of Agriculture. The USDA has a history of gender and racial inequities in its policies and programs.[14] Through USDA programs, the federal government subsidizes agriculture and farmers through a variety of mechanisms including income support, disaster mitigation, crop insurance, and farm loan programs. In 2011, income support payments to farmers were estimated at approximately nine billion dollars (USDA 2012a). Notably, men farmers receive more benefits from commodity payments in conventional agriculture and other government programs than women farmers. This may be due in part to the crops that direct payment programs support, namely wheat, corn, grain sorghum, barley, oats, upland cotton, rice, soybeans, other oilseeds, and peanuts. Farms operated by women have a lower tendency to produce these crops than farms operated by men.[15] In 2007, 45% of male farmers received over ten thousand dollars in government payments compared to just 25% of women farmers (USDA 2009b). USDA loan programs have been criticized for discrimination against women, blacks, and Hispanics, with multiple civil rights suits filed. In 2010, USDA reached a settlement with black farmers of $1.25 billion in a long-standing civil rights suit for black farmers' claims that they were denied USDA loans. In 2011, the USDA established a claims process, promising to bring justice to women and Hispanics by making $1.3 billion available to women or Hispanic farmers who were discriminated against in farm loans between 1981 and 2000 (USDA 2012c).[16]

USDA has instituted other programs recently that assist some women farmers, at least indirectly. In 2008, Congress passed a bill to provide seventy-five million dollars to develop education, training, outreach, and mentoring programs to new and beginning farmers and ranchers after recognizing the increased age of farm operators nationally and the expected decrease in their numbers. Some of these funds are specifically targeted to provide education for women and minority farmers. Another USDA program that indirectly supports women farmers is the Sustainable Agriculture Research and Education Program (SARE). Instituted in 1988, SARE provides grants for agricultural research and education programs that promote profitability; stewardship of the land, air, and water; and quality of life for farmers, ranchers, and their communities. SARE funding has increased over time, and in 2014 the Farm Bill provided $22.7 million to SARE, 18% greater than the previous highest level for the program. SARE supports a wide range of efforts, including sustainable and organic crop production, cover cropping, composting, development of local food systems, and diversity of producers. Since women are more likely to be engaged in sustainable agriculture than in large-scale commodity production, these funds, while not directly targeted to women farmers, support their efforts. In addition, the Risk Management Agency of USDA worked in conjunction with Cooperative Extension to develop and offer Annie's Project, an educational program on farm business management designed specifically for women. Direct competitive funding from these various agencies within the USDA has helped establish women's agricultural networks in various states including Pennsylvania, Vermont, and Maine.

With the exceptions of instances noted above, agricultural institutions remain largely intransigent to shifts toward gender equality and surprisingly entrenched in patriarchal ideologies and practices. This entrenchment continues to enhance conventional farming, reinforcing established barriers for women to enter (or to be represented in) agriculture. However, women have found significant opportunities within alternative agricultural systems, including the organic, sustainable, and local agrifood movements.

Opportunities amid Shifts in Agriculture

Women Farmers Seize Opportunities and Provide Growing Interest in Sustainable and Organic Agriculture and Local Food Systems

The development of alternative approaches is a relatively recent shift in agriculture. Women farmers are actively involved in organic, sustainable, and local foods movements (Chiappe and Flora 1998; DeLind and Ferguson 1999; Hassanein 1999; Meares 1997; Liepins 1998; Trauger 2004; Trauger et al. 2010). These three movements in agriculture have gained increased attention and challenged the broader agro-industrial food system. The organic agricultural movement began as a rejection of the chemical intensive agricultural production system that emerged after World War II. Rather than relying heavily on chemical fertilizer and pesticide use, organic agriculture emphasized building soil quality and health to increase production and was intimately tied to the natural foods movement. In the US, J. I. Rodale and others advocated for organic agriculture in direct criticism of the increasingly intensive use of pesticides in conventional agriculture. Rachel Carson's popular and influential book *Silent Spring*, published in 1962, strongly criticized pesticide-intensive agriculture, especially the use of DDT, for creating serious environmental and human health problems. The critique of the widespread and inappropriate use of pesticides provided clear justification for the practice of organic agriculture.

Organic farmers in the 1970s and 1980s often saw themselves as providing a radically alternative model to conventional agribusiness. By the late 1990s, growing consumer awareness of pesticide residues on food, environmental problems associated with pesticide use, and health concerns with highly processed food resulted in a steadily increasing demand for organic food. Organic agriculture has grown substantially in recent years, with land in organic production increasing from less than a million acres in 1992 to 5.4 million acres in 2011 and with the number of certified organic livestock increasing from 11,647 in 1992 to 492,353 in 2011 (USDA 2013d).[17] Not only are many small-scale farmers, including women, moving into organic production, women farmers are more likely than men farmers to produce organic products (USDA 2014e).

Organic agriculture is no longer the purview of small-scale farmers, however. The growth of the market for organic products has resulted in large-scale producers and distributors in the organic marketplace, and in the process, organic agriculture has been codified, standardized, and corporatized, allowing it to fit more neatly into the broader agrifood system (Guthman 2004). In contrast, many small-scale farmers who do not use chemical fertilizers or pesticides may choose to forego organic certification. They report lacking the time or money to meet the record-keeping requirements and to manage the certification and inspection process. These farmers often refer to themselves as sustainable farmers.

But sustainable agriculture encompasses more than organic agriculture. The concept of sustainability derives from the broader environmental movement's emphasis on sustainable development, which attempts to promote economic growth without compromising the environment or the longer-term health of future generations. Sustainable agriculture challenges conventional agriculture by promoting paradigm shifts that are broader than production practices. The concept of sustainability is often described as a three-legged stool that incorporates environmental, economic, and social sustainability. Sustainable agriculture has been adopted by international development organizations, by alternative farmer movements in the US (such as the Pennsylvania Association for Sustainable Agriculture), and, to a limited degree, by some agencies within the US government (such as the SARE program within the USDA). However, some also worry that the term "sustainable agriculture" has been co-opted by several large agrichemical companies that claim to be supporters of sustainable agriculture.

More recently, perhaps in reaction to the corporate move toward organic agriculture, and the continual struggle for smaller scale farmers to survive, the local foods movement has emerged. Local food systems offer alternatives to conventional agriculture, including greater trust between farmers and consumers (Hinrichs 2000; Jarosz 2000), shorter food supply chains, and reduced social, economic, and geographical distance between farmers and consumers (Wittman, Beckie, and Hergesheimer 2012). The growth in the local food movement involves an increasing number of farmers' markets, direct sales of farm products to consumers and restaurants, and community supported agriculture.[18] Community supported agriculture (CSA) is one of the most prevalent forms of organizing local food systems.[19] Limited consensus exists on criteria for what classifies a farmer as local. A recent study in Washington State found that women outnumbered men as vendors in farmers markets and as farm market managers (Ostrom 2014).

Alternative food systems represented by organic, sustainable, and local agriculture stand in direct contrast to large-scale conventional commercial agriculture. The distinction between conventional and alternative agriculture as described by Beus and Dunlap (1990) involves substantial paradigm differences. The conventional paradigm represents large-scale industrial agriculture and is characterized by centralization, dependence, competition, domination of nature, and specialization. In contrast, alternative agriculture emphasizes ecologically sustainable practices which are characterized by decentralization, independence, community, cooperation with nature, and diversity. These differences are manifest in the size, scale, and practice of sustainable and conventional agriculture. But technology and the growth of alternative agriculture and food systems are not the only determining factor in the type of farming that women choose to practice. The resistance of conventional agricultural institutions to gender equity may also help explain women's interest in alternative agriculture. And in fact, Chiappe and

Flora (1998), in a study of Minnesota farm women committed to sustainable agriculture, found that women's perspectives expanded Beus and Dunlap's framework of alternative agriculture to include the importance of family life and spirituality and a more nuanced concern with community.

The growth of alternative agriculture over the past few decades has created openings for women not previously available. In a similar fashion, women are gaining economic, political, and social power outside of agriculture. Therefore, FAST must explore how gender relations outside agriculture set the stage for more women entering agriculture and sustainable agriculture in particular.

Trends outside of Agriculture That Shape Women's Opportunities

Several concurrent trends in gender equity in the broader US culture, political system, and economy are essential to note. These trends influence women's expectations related to their education, employment opportunities, and roles within the household. Women's increased participation in the labor force, particularly since the 1970s, has resulted in the opening of career opportunities (and related educational preparation) for women not seen earlier. In recent decades, women have increasingly started their own businesses, creating the possibility of women running their own farm-based businesses. At the same time, women have continued to take responsibility for issues related to the home and family, particularly food and childcare. As a result, women tend to express greater concern about issues that touch close to home, such as environmental quality, nutrition, and health. Here we document each of these trends and describe their influence on women in agriculture.

INCREASING GENDER EQUALITY

Gender inequality in the economy, politics, and the family has declined in the US since the 1960s, but it has not been eliminated (Gauchat, Kelly, and Wallace 2012). In the mid- to late twentieth century, the civil rights movement, enactment of laws prohibiting sex discrimination, and advances of the women's rights movement created an atmosphere that was hospitable to more women working outside the home. The combination of these factors created strong incentives for women to join the workforce, and increased their participation rate in historically male-dominated occupations.

In the economic sphere, women have entered the paid labor market at unprecedented rates accompanied by slowly declining gender inequality in earnings and occupational segregation.[20] After World War II, women's participation in the US labor force expanded greatly. Soon after the war, less than one-third of women were in the labor force. By 1999, women reached the peak of their labor force participation, 60%. Since then, however, labor force participation among women has declined slightly. In 2012, 57.7% of women were in the labor force, compared to 70.2% for men.

Despite the great strides made by women in society and in the labor force, statistics reveal a continuing lag in US social and economic trends. Women still earn less and are more likely to live in poverty than men.[21] In part, this is because differences in employment distributions of women and men with women concentrated in lower-paying and traditionally female occupations (White House Council on Women, n.d.; Wootton 1997). Women are still underrepresented in the executive ranks in companies, and they are paid significantly less than men on average. Because women earn less and two-earner households have higher earnings, families headed by women have far lower income than do married-couple families. Single-mother families face particularly high poverty rates, often because of the lower wages earned by women in single-mother families (White House Council on Women, n.d.). Women are still less likely to work in the paid labor force than are men, are still less likely to work part-time, spend more of their time in household activities or caring for other family members, and do more unpaid volunteer work. Analysis of gender inequality in the labor force reveals complex relationships among gender, race, and class (Acker 2006; Collins 2004).

McCall (2001) adds the dimension of local level economic conditions and coins the term "configurations of inequality," which is helpful in understanding women's position in agriculture. McCall's typology of labor markets compares industrial configurations of inequality typical of manufacturing with postindustrial configurations of inequality typical of high-tech jobs. She finds that high levels of gender inequality characterize industrial labor markets. Industrial labor markets that flourished through several decades in the US after World War II were characterized by manufacturing employment, labor unions, and dominance of large corporations that benefitted white male manual workers with high school educations and disadvantaged women and people of color (Gauchat, Kelly, and Wallace 2012). Gains in education and labor market participation of women reduced the advantage of white men in the labor market (England 2010). Postindustrial configurations are more complicated—in some cases, gender inequality is reduced, while in others, gender inequality is reinforced.

Another change in the postindustrial labor market is that workers are no longer expected to work for one employer for their entire careers. This shift occurs as workers seek better job opportunities and corporate or organizational policies make workers vulnerable to job loss (Williams, Muller, and Kilanski 2012). Rather than follow a traditional career path of moving up in an organization, people change jobs and change companies both as a choice and as a necessity. Many women's lives never fit well with a traditional career path given their family responsibilities. Women, especially mothers, have rarely followed the male pattern of continuous, full-time employment for one employer (Stone 2007a). Self-employment or starting a business, including farming, is one strategy some women pursue as they change jobs for various reasons.

While women have entered traditionally male jobs and occupations, the work culture has lagged behind in adjusting to the multiple responsibilities that many women hold. In her study of why high-achieving women opt out of their careers, Stone (2007a) found that women today "choose" to be home full-time not so much because of parenting overload but because of work overload, specifically long hours and the lack of flexible options in their high-status jobs. "The popular media depiction of a return to traditionalism is wrong and misleading. Women are trying to achieve the feminist vision of a fully integrated life combining family and work" (Stone 2007b, 19). The prospect of farming may well be appealing to women who want to balance their time between work and family responsibilities.

INCREASE IN WOMEN-OWNED BUSINESSES

Women's increased involvement in farming parallels a larger trend in the United States of women's increased involvement in business in general. Women-owned firms accounted for 28.7% of all nonfarm businesses in the United States (US Census Bureau 2012). In 2007 women owned 7.8 million nonfarm US businesses, which was an increase of 20% since 2002. The question of why women, and workers in general, have moved toward self-employment has no simple answers. Some argue that women are pulled into self-employment and small business ownership for emancipatory reasons, while others argue that they are pushed into self-employment due to economic restructuring and unemployment (Hughes 2003). Hughes found that most women who are small business owners and those whom she interviewed in Canada started their businesses because they desired a challenge, a positive work environment, independence, and meaningful work. Studies in the US also note that the majority of women start small businesses to fulfill entrepreneurial desires, with only a minority citing job loss as their motivation. The vast majority of women business owners in the US reported they were self-employed to fulfill entrepreneurial desires, with just a minority citing downsizing and explicit gender barriers (Carr 2000). But there is no doubt that there is also a push to find more fulfilling professional lives. Of the women who leave jobs to start their own business, many leave negative work environments marked by highly bureaucratic organizations, lack of independence, and limited decision making opportunities (Hughes 2003).

Although women now own businesses in unprecedented numbers, entrepreneurship continues to be viewed as the province of men, a pattern in conventional agriculture. P. Lewis (2006) argues that the dominant discourse of entrepreneurship focuses on "heroic masculinism." Entrepreneurs are defined as possessing stereotypically masculine qualities such as risk taking, leadership, and rational planning (Bruni, Gherardi, and Poggio 2004). Others have suggested, however, that women entrepreneurs create a different model of business, reframing entrepreneurship. Canadian women who are small business owners did not desire to grow their businesses but were satisfied with their small, stable operations (Lee-Gosselin and Grise 1990). They preferred a model of entrepreneurship in which they maintain control of the operation, preserve a work-family balance, enhance their quality of life, and focus on the customer and employee. Fenwick (2003) also reports that women's businesses offer new possibilities for alternative models of entrepreneurship. Some successful women entrepreneurs downplay gender discrimination or essential characteristics of women as impacting their businesses (P. Lewis 2006). McGregor and Tweed (2002) also question the conclusion that women who are small business owners only want to satisfy their intrinsic needs and are not interested in growth. Indeed, they found variation among women who are small business owners, finding that some women, those who participated in business networks, were more oriented toward growth than other women (and men) business owners.

What is significant is that the desire for self-employment and entrepreneurship among women farmers and the reasons for making this choice are also prevalent in American society among other contemporary women. The growth in the number of women farmers can be seen as part of a larger concurrent trend among women.

One of the paradoxes of modern gender relations is that even as women's labor force participation has increased, the time women spend on traditional household and family duties has not measurably decreased. Women still maintain responsibility for food preparation, childcare, health, and home management. This set of responsibilities makes women more attuned to issues such as food, food quality, health, and environmental quality.

WOMEN'S CONCERN FOR ENVIRONMENTAL QUALITY

Women farmers' involvement in environmentally sustainable agriculture also parallels the greater concern women in general have for environmental issues. Women consistently express more concern about environmental issues (Xiao and McCright 2015; Jones and Dunlap 1992; Goldsmith et al. 2013; Zelezny, Chua, and Aldrich 2000), awareness of environmental risks (Davidson and Freudenburg 1996; Flynn, Slovic, and Mertz 1994), and response to environmental problems (McCright 2010) compared to men. Women's higher level of concern about the environment remains consistent across a wide range of studies and contexts. There is less agreement on the explanation of why this difference exists. Davidson and Freudenburg's classic study (1996) systematically analyzed multiple explanations for these differences. The two most consistently supported explanations for women's higher level of concern for environmental issues are their greater concerns for health and safety, especially when children are present, and their lack of trust in the institutions responsible for managing risks. Davidson and Freudenburg reject the idea that women are more concerned because they have less knowledge about science and technological issues. A study by Flynn et al. (1994) found that white women and both nonwhite men and nonwhite women were more likely to be concerned about a wide range of environmental risks because they are less likely than white men to benefit from technologies and institutions and because they assert less power and control in these institutions. These studies eliminate support for biological and essentialist explanations for women's concern with environmental risks wherein women are assumed to be closer to nature, and point to social and economic explanations related to the roles that women perform in families, households, and communities. Zelezny, Chua, and Aldrich's (2000) study of women and men in fourteen countries found that women report stronger environmental attitudes and behaviors than men and that women's greater concern for the environment is because women are more likely to consider the perspective of others and are socialized to have an ethics of care where they are more likely to consider the perspective of others than are men who are socialized to value independence and competition over the consideration of others. An alternative explanation is offered by a more recent study by Goldsmith et al. (2013). They argue that men are less likely than women to admit and tackle environmental problems because men benefit from social, economic, and political institutions and thus have stronger motives to justify the status quo.

In contrast, ecofeminists argue that women are more concerned about the environment due to the connections and closeness of women to nature. Merchant's (1980) classic work *The Death of Nature* painstakingly traces how science in the seventeenth century set the stage for ecological destruction that relied on the connection between the domination of nature and the domination of women. Soon after Merchant's work was published, ecofeminism emerged as a social movement. Ecofeminists recognized the connections between the domination of nature and the domination of women. Some ecofeminists celebrated women's biological connection to nature through childbirth as well as their role in nurturing, including their role in providing and preparing food. Ecofeminism served as a rallying cry for women to become more involved in the environmental movement.

In the 1980s, ecofeminists were met with criticism from academic feminists for their essentialism in connecting women with nature and valorizing women's role as nurturers. Also, the implication of ecofeminism indirectly suggested that women should be responsible for cleaning up environmental problems men created and that men were not responsible or capable of having a close relationship with nature. By identifying women with nature and men with culture, ecofeminists were cementing their own oppression (Merchant 2006). Other feminists concerned with the environment and many ecofeminists took the criticism of essentialism seriously and began crafting new theories to explain the connections between women, gender, and the environment. Feminist political ecology rejected biological essentialist notions of women's connection to nature, emphasizing that women's concern for the environment is connected to their daily activities, labor, and access to resources (Rocheleau 1996; Agarwal 1992). Much of the scholarship on feminist political ecology focuses on the daily lives of rural women in developing countries, but it provides a helpful framework to understand women's commitment to sustainable and organic agriculture in the US.

Women's concern with the environment in agriculture is perhaps best personified by the work of Rachel Carson, a scientist and writer concerned with the uncritical and widespread (and often ineffective) use of broad-spectrum insecticides. Even though we have become more aware of the risks associated with pollutants in the environment since the publication of *Silent Spring* (1962), we now apply more pesticides. Now, as then, the federal government establishes maximum permissible limits of contamination, called "tolerances" (US EPA 2012) for individual pesticide residues in food. Carson considered the US system of allowing certain levels ("tolerances") of pesticide in foods, and the resultant innumerable, recurrent, small-scale exposures, more important than contamination from mass spraying. She also called attention to the development of resistance to commonly used insecticides and pest resurgence due to the destruction of natural enemies of insects. These problems not only continue, but have increased today in crops grown using conventional pesticides (Li, Schuler, and Berenbaum 2007; Dutcher 2007) and in transgenic crops (Dill 2005; Tabashnik et al. 2008; Gassman et al. 2014).

When *Silent Spring* was published in 1962, representatives of the agricultural and chemical industry, government officials, and scientists reacted by branding Carson as an emotional spinster alarmist, a quack, a Communist, and worse (Lear 1997). On the contrary, in *Silent Spring*, Carson recognized the need to manage insect pests of agricultural and public health significance. She did not argue for an end to pest control, or even to the use of chemicals for pest control. Rather, based on a review and synthesis of scientific literature and media reports, she solicited an end to the widespread use of broad-spectrum insecticides, mainly organochlorines and organophosphates, and sought the development and adoption of less toxic chemicals and more narrowly focused alternatives based on the best scientific and ecological knowledge. As an alternative to the use of broad-spectrum insecticides, she offers approaches that use narrower spectrum, less persistent chemicals and nonchemical methods such as biological control of pests through the use of the natural enemies and pathogens of insects, and where appropriate, sterile insect control techniques. These are approaches that are increasing today, but the dominant approach to pest management remains chemical control. Carson argued that chemicals should not be used without adequate scientific and public knowledge about and consideration of their impact on the environment and human and wildlife health. We are the beneficiaries of a lasting legacy of Rachel Carson and her prescient analysis of pesticide intensive agriculture. She had a tremendous influence on US national pesticide policy that resulted in a ban on DDT and other pesticides in the US and inspired an environmental movement that led to the creation of the US Environmental Protection Agency (J. Lewis 1985). Women in sustainable agriculture continue the legacy of Carson's work by acting to change our dependence on chemicals with new agricultural practices and technology.

Closer to the Table: Women's Continued Responsibilities for Food

Women in the US do the majority of household food-related work (Bowen, Elliott, and Brenton 2014). Food production may still be male-dominated, but the closer food moves from farm to table, women have more responsibility for food work in acquiring, processing, cooking, and serving food (Allen and Sachs 2007). In their households, women often spend substantial amounts of time and energy to maintain nutritional and sanitary standards that involve hours of shopping, meal planning, preparation, cooking, washing dishes, and cleaning kitchens. Food work requires physical labor but also entails mental and caring labor such as planning meals, incorporating nutrition, knowing the food preferences of family members, planning the timing and location of meals, keeping up with news on nutrition and food safety, and arranging and serving meals (Allen and Sachs 2007; DeVault 1991). Food provision is essential to many women's identities and gives them power in their households and in their communities (Counihan 2004; Devasahayam 2005). But as Allen and Sachs argue, this food-related work is also a key component of both their exploitation and their resistance. Disagreements continue over "whether women's food work gives them power in the family or reinscribes their subordinate gender roles" (Allen and Sachs 2007). Either way, women's work with food ties them to their families and friends and in some cases maintains cultural traditions that are at the heart of many women's identities.

The increased employment of women outside the home has also led to the movement of food-related work, such as processing, cooking, and serving food outside the home, including eating in restaurants as well as purchasing prepared foods from grocery stores. In the US, there has been a continual shift in the share of the national food budget spent on food consumption away from home, from 9% in 1900 to 49% in 2008 as the resources of women increase (Schnepf and Richardson 2009, 9).

The recent local foods movements[22] decry reliance on commodified, industrialized food in favor of eating locally grown and less processed food (Martinez et al. 2010). The move to consuming less processed, local foods implies that someone has to prepare and process these local foods and most often this someone is a woman. In households that subscribe to CSAs, women tend to be more involved than men: in their study of CSAs, Cone and Myhre (2000) found that 81% of women picked up vegetables compared to 43% of men, 88% of women processed vegetables compared to 34% of men, and 88% of women cooked for their households compared to 44% of men (192). The shift to local food might foretell women returning to their home kitchen processing food, cooking, and serving others, but Shannon Hayes (2010), in her book *Radical Homemakers*, rejects this assumption. Putting less faith in the market and decrying consumer culture, Hayes calls for women (and men, too) to become radical homemakers and to reclaim domesticity. She gives concrete examples of women who have reclaimed a meaningful and sustainable domestic life without inadvertently reinscribing their own subjugation. She begins her book by explaining why it is possible for a feminist to can tomatoes, but she goes much further in calling for a rejection of consumer culture followed by a reframing of daily life, households, and communities in the US to more self-reliant and community-based livelihoods. Women have stepped into the local food scene as entrepreneurs and farmers. They now produce food for local consumption, start food processing operations, and cook in up-scale restaurants.

In sum, what has occurred among women outside agriculture in their choice of entrepreneurship and labor force participation, their concern and action for the environment, and their involvement closer to the table not only illuminates, but indeed validates, the choices that the women in alternative agriculture have made because the choices permeate our society as a whole. The choices are not unique to the women farmers. Further, we suggest that what has occurred among women outside agriculture has laid a foundation for the women in alternative agriculture.

In the following chapters we examine some of the challenges that women in the Northeast face as they pursue a life in farming, and how some have faced these challenges by forging a new type of agrifood system. We seek to address a number of questions, framed from a feminist perspective, about why women don't have as equal access to resources and opportunities in agriculture as men and about why fundamental justice in agriculture related to women and gender is lacking. Our work draws on these feminist principles to guide our grounded theory to better serve the interests of women farmers by exposing issues of gender and power that influence their experiences. Our knowledge has been developed through a connection between praxis (working with women farmers in the Northeast) and research (both conventional and participatory). These questions encapsulate over ten years of quantitative and qualitative data generated through multiple modes of collective inquiry and experimentation grounded in experience (appendix) to culminate in our proposition of FAST.

In chapter 2, we ask, "How do women farmers in the Northeast view themselves? How do others view them? What are the barriers to claiming the identity of 'farmer'?" Through survey and interview data, women farmers reveal how they develop, perceive, and arrive at their identities in relation to the work they do on the farm, and they describe some of the challenges in being recognized as "real" farmers by family members, other farmers, and agriculture-related service providers. We describe how several factors influence women farmer's identities and how agricultural institutions shape women farmers and are being shaped by them.

In chapter 3, we ask, "How have women gained access to resources in farming in patriarchal systems that favor males accessing the traditional resources in farming?" Women farmers describe a number of the challenges in their efforts to secure the basic resources of farming—land, labor, and capital—and explain how they have addressed these challenges by employing innovative strategies as alternatives to the traditional means of obtaining these resources. For example, some women farmers draw on the growing interest in agriculture among the nonfarming populace to secure labor, and they take innovative approaches to manual work aided by devising, adapting, and using appropriate tools and equipment. To obtain resources to finance land and support for their operations, some women farmers are drawing on emerging services and programs that specifically target women farmers or the type of farming that they do.

In chapter 4, we ask, "How do women in agriculture affect the larger agrifood system? What choices do they make, and how do their choices in farming influence their communities and environment?" Women farmers in the Northeast describe multiple goals that they have for their farms. Their goals reflect personal commitments that are oriented toward both individual and local values, as well as community and global concerns. These women describe several innovations that they employ on their farms, such as new products, innovative enterprises, and business models, along with an emphasis on multiple aspects of sustainability to achieve these goals, all of which serve to create a new food system one farm at a time.

In chapter 5, we ask, "To what extent do existing educational programs and resources respond to the needs of women farmers, and do organizations that typically deliver this type of information recognize women as farmers with a need for a mix of general information and information tailored specifically for them as women farmers? How do agricultural organizations create spaces for women's voices in agriculture? How have women farmers responded to agricultural organizations' lack of an adequate response to their educational and social needs?" Women farmers articulate their educational needs—content, format, and context. We compare these stated needs to the opinions of existing providers of agricultural information, in particular Cooperative Extension, a primary source of educational programs and technical assistance for farmers. We highlight four instances where women have created agricultural networks (including the Pennsylvania Women's Agricultural Network, PA-WAgN) in response to a lack of adequate organizational support. Women farmers developed networks to intentionally focus on accessing the resources they need to successfully overcome the challenges that all farmers face, as well as those they face as women. In addition to providing access to knowledge and skills, these networks provide access to new ideas from other women farmers and social support that can alleviate isolation, legitimate women's identities as farmers, increase their capacity as farmers, and as a result, enhance their farm businesses.

Finally, in chapter 6, we introduce a new feminist theory of agrifood systems (FAST) based on our research and experiences with women farmers. FAST can provide a context for comparing and understanding the differences and similarities in the role and influence of women farmers nationally and internationally. We argue that women farmers, through their individual and collective efforts, are transforming Devine's (2013) theory of agrarian feminism by providing both a critique of and an alternative to the conventional and patriarchal system of agriculture. This new theory is characterized by five interrelated themes, in which women farmers (1) create gender equality on farms, (2) increasingly assert the identity of the farmer, (3) access the resources they need to farm, including finding innovative ways to access land and capital, (4) shape new food and farming systems, (5) navigate older agricultural organizations and institutions that have not always met their needs, and (6) form new organizations for women farmers. These six themes, which we will discuss in detail in the intervening chapters, form the basis of FAST, which provides a framework for understanding the shifts in agriculture and women's roles in agriculture in the Northeast.

Chapter 2

Tilling the Soil for Change: Claiming the Farmer Identity

Natalie Ingram and her husband produce organic grain crops on more than 1,300 acres. Even though Natalie considers herself a farmer, she is concerned that within the agricultural community, the women who farm with their husbands are not seen as *being* farmers. She describes attending numerous agricultural meetings where men are perceived as farmers and women are viewed as "just the farmer's wife." Shortly before her interview with us, she had read a report from a recent pasture management meeting that she had attended. She was livid at how women were portrayed and expressed her frustration: "Kathy [a fellow woman farmer and friend] was identified as 'wife of a dairy farmer.' Of all of the people who do not deserve to be identified as 'wife of,' Kathy is more of a farmer than her husband is. To me that kind of social stereotyping just because a woman stands up and says something or is involved in some way, she is automatically being classified as wife of the farmer. Someone should have known better." Natalie considers describing women who are actively farming as farm wives belittles their important contributions to the farm operation. In addition, she points out that she expected a more accurate portrayal of Kathy's work on the farm from the report's writer and editor who chose to refer to Kathy as a "wife of a dairy farmer," rather than a "dairy farmer."

The lack of progress in how women's work on farms is perceived and labeled by men reflects the way women farmers are embedded in an agriculture still characterized by patriarchal norms. However, as Natalie's comments above reveal, women are now identifying their role as a farmer, even within farming partnerships based on heterosexual marriage. This shift stands in contrast to Devine's description of agrarian feminism characteristic of previous generations of women farmers who placed a higher value on their dependent role rather than on their independent role as women in agriculture.

Concurrently, many women who farm with their husbands, other partners, or on their own struggle to be taken seriously as farmers by other farmers in their communities; by institutions that have traditionally supported farmers, such as agribusinesses, lenders, and Cooperative Extension; and even by their families. A lack of recognition as farmers by others can lead to challenges for women farmers, such as isolation and self-doubt, as they negotiate their identities (Trauger 2004; Trauger et al. 2008, 2010). Nevertheless, it can also lead them to develop new and different ways of working on the farm and in the community that support their identities as farmers.

Social psychological theories describe how a person's identity develops in interaction with others, such as family members, colleagues, and community members. Identity is built by performing tasks consistent with a specific role, such as an occupation (e.g., teacher) or family member (e.g., mother). Most people fulfill multiple roles and adopt a role based on the context and the expectations they perceive others to have for them in that context (Stets 2006; Stryker 1980). Thus, for a farming woman to label herself as a "farmer" means to perform the tasks associated with that role and have her role confirmed and supported by other relevant individuals and groups. To the extent that she is limited in her ability to perform that role, or receives positive or negative feedback from others about her performance of that role, she may reexamine whether she should have "farmer" as part of her identity (Brasier et al. 2014; Burton and Wilson 2006; Stets 2006).

Women farmers' identities provide an important lens through which we can begin to understand their personal struggles. In this chapter we use our data from surveys and interviews of northeastern US farmers to describe how women who farm or live on farms identify themselves.[1] We examine the ways in which women farmers describe challenges in identifying and being recognized as real farmers, as opposed to being dismissed as hobby farmers. Through our survey and interview data, these farmers reveal how they develop, perceive, and arrive at their identities in relation to the work they do on the farm and how this identification process affects the use of their bodies in farming. These data also illustrate the ways in which these women farmers perceive the support (or lack of) from family members, other farmers, and agriculture-related service providers. The pathways by which women enter farming, for example, by buying a farm outright, through inheritance, marriage, or other partnership can also affect how these women through farming can shape the development of farm women's identities. A focus on how these factors influence those identities can help us understand how broader institutional contexts shape and, in turn, can be shaped by women farmers.

The Changing Definitions of "Farmer"

The share of US farms with women as the principal operator nearly tripled over the past three decades, from 5% in 1978 to 14% by 2007 (Hoppe and Korb 2013). As of 2007, women comprised 30% of all farm operators in the US, a 17% increase from 2002 to 2007 (USDA 2009b). Part of this increase is due to new counting procedures at the USDA that uncovered that more women are farming than was officially tallied in the past. Before 2007, only one operator was counted per farm, and usually the male operator was counted. Tallying more than one operator per farm substantially increased the number of women who were officially counted as farmers. But the increase in the official number of women farmers is not solely due to new methods of counting farmers; more women are actually entering farming as the principal operators of their farms. The number of women principal farm operators increased from 237,819 in 2002 to 306,209 in 2007 for a 29% increase in women farmers in just five years (USDA 2009b). This increase in women farmers likely represents changes in both the number of women entering farming and the number of women identifying themselves as farmers.

Historically, strong identities of men as farmers and women as farm wives typified farms in the US. These gendered identities often coincided with divisions of labor on farms; however, many women were and are heavily involved in multiple aspects of farm operations that transcend the gendered divisions (Barbercheck et al. 2009; Sachs 1993). And farming is not the single economic activity for most farm households. Farm operators and their spouses or partners often allocate time to off-farm work activities, either working for wages or salaries or operating a nonfarm business. Depending on the type of farm, the share of total household income derived from off-farm sources ranges from 14% to 93% (USDA 2012a). Given the level of off-farm work performed by US farmers, farm women are often necessarily an important source of labor and management on farms. Studies in the 1990s found that women identified with the role of farm wife even if they participated extensively in the farm operation (Whatmore 1991; Sachs 1993). Some women may choose to identify as farm wife, as opposed to farmer, because they may benefit from the economic security, respectability, and prestige associated with that title in rural communities (Fink 1992).

However, the rigid identities of men as farmers and women as farm wives are beginning to break down. Brandth (2002), in her study of changes on family farms in Europe, suggests that "the way we speak about men and women on farms may not correspond to the way they really are" (182). Many women no longer identify only as farm wives but as multiple roles, including farmer, housewife, agricultural worker, off-farm worker, farm assistant, and farm partner, thus reflecting the complex, heterogeneous, and dynamic roles of women on farms. Many farm women recognize these complexities as they describe their identities (Trauger 2004; Trauger et al. 2008).

Audrey Gay Rodgers with prized Ayrshire Plum Bottom Paragon's Petoria. Photograph by Ann Stone, PA-WAgN.

How Do Women on Farms Describe Themselves?

In order to understand these identity complexities, we examine the ways in which women on farms in the northeastern US describe themselves, drawing on both quantitative survey and qualitative interview data (SNEWF and IIWF).[2] Previous surveys of farm women asked them to choose one role such as farmer, farm wife, bookkeeper, or not involved in farming (Rosenfeld 1985; Bokemeier and Garkovich 1987). Rather than follow the example of previous studies that asked for an exclusive role, we designed a survey to allow women to rank themselves more inclusively on seven farm-related roles, including farmer/operator, farm wife/domestic partner, farm bookkeeper, farm business partner, off-farm worker/professional, and farmworker/apprentice (Brasier et al. 2014). For each role, women were asked to rate themselves on two scales by circling a number from 1 to 7. One scale ranged from (1) "does not describe me" to (7) "describes me," and the other ranged from (1) "not central to who I am" to (7) "central to who I am."

For most of the roles, the distributions of the responses of women farmers in the Northeast are bimodal, meaning that most women answered either 1 or 7 (fig. 2). This suggests that women farmers are relatively clear about the roles they perform in relation to the farm and the meaning of those roles to their own identities.

Two roles—farmer/operator and farm wife/domestic partner—are descriptive of the largest percentages of respondents. Nearly two-thirds (63%) selected a value of 6 or 7 for the farmer/operator role, as did 62% for the farm wife/domestic partner role. The roles describing the next largest percentages of respondents are farm bookkeeper (58% selecting a 6 or 7), farm business partner (54% selecting 6 or 7), farm entrepreneur (53% selecting 6 or 7), off-farm worker/professional (53% selecting a 6 or 7), and farmworker/apprentice (33% selecting 6 or 7). The roles with the highest percentages of women indicating that the role is "central to who I am" are again farmer/operator (53% selecting a 6 or 7) and farm wife/domestic partner (52% selecting a 6 or 7), followed by farm business partner (43% selecting a 6 or 7), farm entrepreneur (42% selecting a 6 or 7), off-farm worker/professional (35% selecting a 6 or 7), farm bookkeeper (31% selecting a 6 or 7), and farmworker/apprentice (23% selecting a 6 or 7).

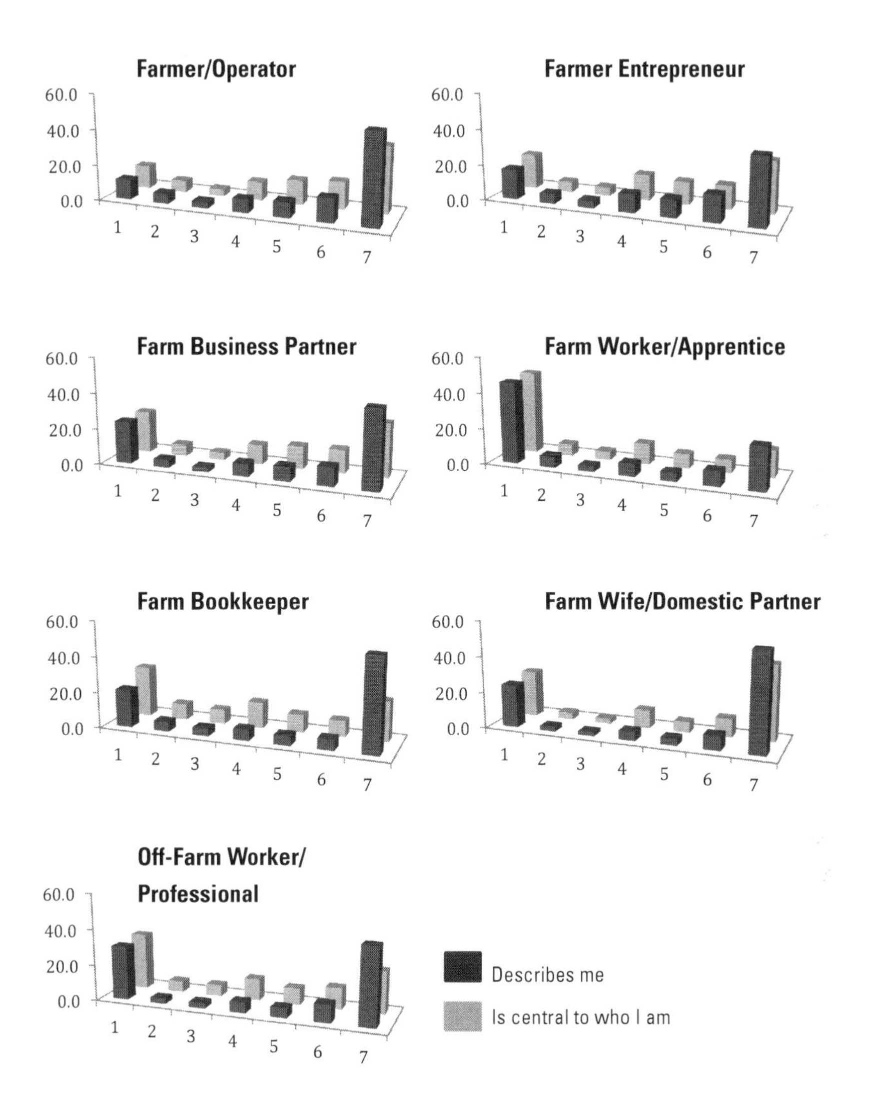

Figure 2. Percentage (vertical axis) of northeastern US farmers responding to a survey question about their role (dark gray bars) on the farm and the centrality of that role to their identity (light gray bars) on a scale of 1 to 7 (horizontal axis). The "describes me" scale ranged from (1) "does not describe me" to (7) "describes me," and the "central to who I am" scale ranged from (1) "not central to who I am" to (7) "central to who I am." Adapted from Brasier et al. (2014).

These data demonstrate that women fill multiple roles on their farms, including keeping books, developing entrepreneurial enterprises, and working on and off the farm. They also indicate that many women perform these roles simultaneously, with both farmer/operator and farm wife/domestic partner as the most prominent roles in terms of both description and importance to them. There is relative congruence between how women rank these roles in terms of how they describe them and how important the roles are. The one exception is the role of farm bookkeeper; while many women are farm bookkeepers, it is not central to the women's identities. This is an important finding because many educational programs geared toward women focus on their roles as farm bookkeepers. Although this may be practically beneficial, our data suggest that focusing on bookkeeping is a missed opportunity to connect with the more prominent identities of women on farms.

In further analysis, we examined the extent to which farm tasks, decision-making authority, and farm and personal characteristics were related to the identities of farm women in the northeastern US (Brasier et al. 2014). This research uses the farm-related roles described above to create two primary identities, farmer/operator and farm partners. The analyses indicate that women who see themselves as farmers/operators are more likely to perform more of the production and business management tasks and to have greater decision-making authority on the farm, whereas women who identify themselves as farm partners are less likely to perform the full range of tasks and are more likely to share authority over farm decisions with someone else. The type of farm (in terms of products), the size of farm, their marital status, and how the farm was acquired were the major factors that explained the differences among women in their adherence to each of these farm identities (Brasier et al. 2014). From this research, we conclude that the two main pathways into agriculture—marrying into a farming family or becoming a farmer on one's own—lead to differences in the type of women's farms, the tasks they perform, their decision-making authority, and their personal identities as farming women.

Clearly, women's identities as farmers develop over time and in concert with their personal histories. They practice these identities through the roles they perform and the tasks that define these roles, and then these identities are reinforced through feedback from others. We next turn to qualitative interview data for a deeper understanding of how women on farms describe themselves, the roles they play, and how their identities have been shaped by their personal histories and their relationships with important other individuals and groups, including family members and the broader agricultural community.

How Farming Women Describe Themselves

Marilyn proudly asserts: "I am a farmer. . . . I have chickens, beef, dairy heifers, pigs, and I direct market all of my stuff." Nan also describes herself as a farmer. To her, being a farmer means not only the physical labor associated with production but also coordinating those activities with the workers and other managers on the farm to generate a product: "The way we work our farm now . . . it has grown a lot, and there are a lot of people working on it, but everybody pretty much has particular crops that they take responsibility for. I am responsible for heirloom tomatoes. What it means to be responsible for it is that. . . . you do not have to do all of the work on the crop, but if the crop needs to be weeded or if the crop needs to be staked or if it is time to pick it, then you have to be sure to inform everybody of what is going on so that we can get that done. One thing that has happened over the years is eventually I have become the supervisor of the greenhouse and I do almost all of the greenhouse planting . . . grow[ing] all of our plants for the field. So we start in about February to grow . . . things." Marilyn's and Nan's identities as farmers are clearly tied to their operations and the work they do on their farms. (But Marilyn reflects more thinking about this, as shown below.)

Being a farmer for some women means being mentally aware of and taking responsibility for a myriad of tasks and roles that taken together, contribute to the success of the farm. As Barb Hartle describes, "Being a farmer, you have to be everything . . . you have to be able to be the marketing manager, and you have to be the distribution manager. You have to be the accountant and the bookkeeper and everything. You are a mechanic, and you have so many hats to wear." For Marilyn, it is the integration of all these tasks and the roles associated with them that define being a farmer but that also provide the reward: "It is very challenging, and it also uses your whole self. You have to make decisions all of the time that involve ecological concepts and financial concepts and mechanical concepts, which I am not good at, but it is good to try and integrate all of it. . . . You have to use your whole creativity and your whole body and your whole self and your whole energy. Most other jobs compartmentalize you in ways. There are not immediate consequences in what you do in a job that is in an office. Where in farming you live with your consequences, and I like that. You have agency in your own work. So that is what I like about it. It is very challenging, but very rewarding." Women who identified as farmers expressed pride and satisfaction as they negotiated what many considered a demanding, complex job.

Pathways into Farming

Regardless of how women enter farming, they define themselves as farmers because of the type of work and decision-making they do to produce and sell agricultural products, whether they farm alone or with a partner. Two main pathways into agriculture for women are through marriage or farm acquisition. Traditionally, men tend to become farmers through inheritance (Alsgaard 2013). Compared to men, women are less likely to inherit a farm but instead are more likely to marry into a farm family. The other pathway for a woman is to acquire (buy or lease) a farm outright, perhaps with a partner. Marilyn chose this pathway. Marilyn wanted to become a farmer from a young age: "I was always interested in farming. . . . I never had the chance to farm growing up. . . . We lived all over the world and lived in apartments and houses in suburbs and things like that. So it took me a while to be able to get to farming." After many years working on other farms and renting land, she has gradually managed to buy a farm and build her farm business. Once she obtained her own land and started producing crops and livestock, she saw herself as a farmer. Marilyn's pathway into farming was decidedly and consciously independent:

> I was not going to marry a farmer to get land. I mean that was not the way I wanted to do it. I saw that happen, but I did not want to be a farm wife. I was a farmer. I wanted to do it and not that "farm wives" do not do it, they just do not identify that way. I was not even about to identify that way [as a farm wife] because I saw it for what it was, which was a sham really. You know the farm was really both people, and everybody in that family knew it. But societally speaking, the guy was the farmer and the woman was the wife. . . . That was not for me, [even though] . . . I did not have the money and no land in the family.

Women such as Marilyn identify their independence as a characteristic that separates them from other women on farms. She takes pride in her independence and lack of reliance on marriage and family to obtain a farm. For Marilyn, being a farmer results in a status that gives her better control and more respect in her community. She believes that although women who identify as farmers and women who identify as farm wives might do similar work, the status as an independent farmer is different.

Women who marry into a farming family may be less comfortable identifying themselves as a farmer. Leisha's farm has a maple syrup operation and a farm camp, and she has identified more as a teacher than a farmer. Leisha married a farmer who recently passed away. Her husband inherited the farm from his parents, who are still alive and participate to a limited extent in the farm operation. She struggles to describe her identity. As she explains, "Some days I think of myself as a farmer, and other days I don't. So I guess I married into farming basically. I never, never had any inkling that I would ever be on a farm in my younger years." Women's identity as farmers is connected to their level of involvement in the farm operation but also to the patterns by which they enter farming. Women who farm independently clearly identify as farmers. Women who farm with their husbands' families or in more complex family arrangements may well participate in the farm operation but may identify simultaneously as farmers and farm partners while others may be hesitant to identify as farmers in their own right.

Women farming with partners identify as farmers even when they divide the labor with others working on the farm. For Barb, farming with her husband eases the pressure of negotiating the multiple roles of farming: "So when you have two people you can split up those roles, and it makes life easier. You can support each other, and you do not have to be unsure about something. You are not the sole decision maker, and you know all of those things, but I think that the biggest one is being able to split up those responsibilities and work as a team." Barb and her husband do all of their major farm planning in the winter and work as a team to decide what they are going to plant, how much they are going to plant, and what seeds and varieties they will purchase. As she says, "That we do together completely." So while Barb identifies herself as farmer, she also sees herself as a farm partner. Many women identify as both a farmer and a farm partner—these identities are not mutually exclusive and break the long-standing perceptions of men as farmers and women as farm wives.

Challenges to Identifying as a Farmer: Family Responsibilities

Farming women frequently juggle familial roles with farming roles. Integrating farm and family responsibilities can result in women taking a less active role on the farm in certain stages of their lives, but it does not necessarily mean they do not see themselves as farmers. Nan identifies herself as a farmer, although her involvement in farming has changed over time in relation to her family responsibilities. When they started their vegetable farm, Nan and Ken did not have children, and they shared most of the work of production and marketing. But as she explains, things changed once they had children:

> I still was involved in fieldwork, and I still was always involved in marketing, but every year as the kids grew up, it changed a little bit. It depended on how much attention they needed or how much time I felt I had to do it. The biggest thing that changes is that the crops became secondary to the children because it used to be that nothing was more important than the tomatoes. Once kids are there, I could legitimately say, "Well it is more important that I get breakfast for the kids than that I get the tomatoes picked." I still was doing it, but it changed my perception of what was important and what was not. I mean before they came, it was sometimes completely overwhelming to me how much pressure the vegetables would put on us. I remember many nights . . . packing tomatoes until ten or eleven o'clock at night.

Nan seems to take for granted that taking care of the children was primarily her responsibility. Caring for children took priority over farmwork, but her identity as a farmer remains unchanged. Many women find their responsibilities on the farm shift when they have young children; they become responsible for, and must balance, identities associated with both productive and reproductive labor. As women take on the productive role in farming in addition to the reproductive role in their families, both roles become part of the farmer identity. This may be especially true for women in agriculture where their productive labor is so closely tied to their reproductive labor, both physically and symbolically. As we detail more thoroughly in chapter 4, the perspective of being a parent aligns many women with sustainable agriculture, reflecting the multiple values women have, including health, environment, and sustainability.

Challenges to Identifying as a Farmer: Women's Bodies and Farm Technology

Women's physical ability to do farmwork is critical to performing the tasks associated with being a farmer. However, many women find using existing farm machinery and related technology challenging given their smaller size and physical strength, on average, compared with men. Further, research suggests that women are more likely to be considered farmers by others if they use tractors and other agricultural machinery than if they do not (Brandth 2006). Therefore, women's bodies, their physical strength, and their ability to use agricultural technology shape their identities as farmers.

Theoretical discourses of farming and farmwork revolve around images of masculine working bodies (Brandth 2006). Cultural constructions of rural masculinity emphasize that these men do hard, physical, dangerous, and dirty work (Peter et al. 2000; Brandth and Haugen 2011; Trauger et al. 2008). Women's bodies are seen as lacking in masculine characteristics, such as strength and endurance, that enable them to farm. Thus, while women may work on farms, their work is viewed as physically easier and therefore of less value than the work of men (Saugeres 2002). Agricultural technology and machinery eases much of the physical labor required in agriculture. Therefore, we might expect a decline in male bodily advantage as farming becomes more mechanized. But the concept of masculinity in farming now incorporates technical competence with farm machinery. Farm machinery is viewed as an extension of the male body, and farmers' desire for larger, more powerful equipment is connected to male bodily strength and competence (Brandth 2006).

Some women farmers express frustration about the physical strength needed to use farm equipment. As Kendra illustrates, "I do a lot of tractor work. Changing the equipment on the back of the tractor is a problem and one of my most frustrating experiences in all of this. It is just because . . . it takes just brute force. Sometimes you just have to hit things with a mallet or kick them. That is not my general way of doing things. So then a lot of times

Participant at PA-WAgN tractor safety, maintenance, and operation workshop at Blackberry Meadows Farm. Photograph by Ann Stone, PA-WAgN.

I feel like it is too hard for me to be doing it by myself. . . . So that kind of thing is a little frustrating to me as a woman, but I think a lot of times men are more willing to just kind of kick things in place." Attaching equipment such as plows to tractors usually involves a three-point hitch, which requires physical strength to accomplish. Other implements are attached to the power take off (PTO) on the tractor, which also requires physical strength and is also dangerous and one of the main culprits in farm injuries.[3]

Women often do not learn mechanical skills, whether they grow up on a farm or not (Leckie 1996). Haley, who did grow up on a farm, explains that she did not learn mechanical skills in school: "There is this tremendous learning curve there because as I mentioned I was not allowed to take shop. Not that I was really interested in taking shop, but had we been forced to . . . I think I would be a little farther along in that type of learning." As Haley's experience shows, the gendered division of labor on US farms often denies women access to important agricultural resources, including information. Women are often not socialized or educated in mechanical skills, which can exacerbate physical differences between men and women. Leckie (1996) posits that this exclusion comes about through the ongoing social processes of agrarian patriarchal culture, operating both inside and outside of the family unit.

Other women that we spoke to also expressed frustration with farm equipment. Although they know how to operate equipment, they would rather not. Nan says, "[We] were both pretty much equal partners in the way that we did the equipment work. I did a lot of disking and I did a lot of mowing and I did a little bit of plowing, but not too much cultivating. We had several tractors then. We had two tractors. I found that I really did not like the equipment work that much, mainly because it is noisy and it is smelly sometimes. I liked doing a little bit of it, but I really did not want get too involved in all of the equipment part." Other women express frustration with farm equipment because it does not fit their bodies. As Kendra explains, "I have been frustrated with not-to-women scale of equipment." She gives the example of a walk-behind tiller. "The size of hands you need in order to stop the thing. You know to get your hand around. . . . You have to use two hands basically." Women's frustration with machinery reflects both the machinery itself and the gendered socialization of men and women with machinery.

Diane Wechter participating in a PA-WAgN tractor operation workshop. Photograph by Ann Stone, PA-WAgN.

Women farmers must negotiate their physical strength and mechanical competence in their farm operations, frequently finding ways to innovate. We find that women farmers often adopt new types of production strategies that require less physical strength and less reliance on large equipment (Trauger et al. 2008). Others see the lack of farm machinery that fit women's bodies as an entrepreneurial opportunity. Ann Adams and Liz Brensinger recognized that the tools they were using to farm and garden did not fit their bodies, so they started a business designing and selling tools that would fit the bodies of women (Kivirist 2012). They conducted research with women farmers and found they were dissatisfied with their ability to efficiently use hand tools. The first tool they designed was a shovel that is lighter in weight than a typical shovel and modified to take advantage of women's lower-body strength. They also found that women farmers reported trouble with larger scale equipment, especially when attaching implements to a tractor with

Tianna DuPont demonstrating the proper way to use a broadfork at a PA-WAgN workshop. Photograph by Ann Stone, PA-WAgN.

the standard three-point hitch. Women they interviewed explained that the three-point hitch has a "flawed design," is "terrible," is "too heavy," and is "too difficult to attach without help." In response, they developed a rapid hitch that does not require the heavy lifting to attach implements to tractors.[4] Clearly, women's bodies and physical strength, in addition to disadvantages in mechanical skills education, are challenges women face as they farm. Many women innovate by using different types of equipment, creating user-friendly infrastructures, and choosing farming methods and types of farms that require less physical strength. Notably, these women adapt so they can claim the identity of farmer even though their ability to use technology related to being a farmer presents many challenges. This work is part of the process of constructing identity.

Challenges to Identifying as a Farmer: Stereotypes of "Farm Wives"

For women with partners, a gendered division of labor can contribute to differences in perceptions of who is "the farmer." Nan and Ken both consider themselves farmers but maintain a clear division of labor on their farm. Nan thinks that perhaps because she does different types of activities on the farm than her husband that people consider it his farm, even though she also owns and works on the farm. Nan resents that her husband and workers on the farm often call the farm "[Ken]'s farm." As she explains, "Sometimes I resent it when he [Ken] does it. I resent it when he says 'Well, my farm. . . .' I said, 'It is not your farm. . . .' I resent it when people that I work with . . . call it Ken's farm. I mean, that is ridiculous, and I do resent that. . . . I am obviously a part of it . . . whether I am doing exactly what they think I should be doing or if they think that is what farmer is supposed to do or that I am where I am supposed to be."

Marilyn thinks the stereotype of the farm wife makes fitting into a rural community easier for married women who are farmers than for single women who are farming. "Married women farmers fit in the community, quote, unquote, 'better.' You fit people's images of what a farmer is. You just fit better. You do not have to push yourself as far as you have to if you are a single woman farmer. . . . The woman that is doing all of the farming. You have to go into the sale barn and deal with the guys and do the stuff, and they do not know you are somebody's wife, and if you were, it would be the guy that is doing it most times. It is just real different. You have to just say, 'Okay, here we go.'" She thinks the women in the community who are part of a farm couple are treated differently than women who are farming on their own, although she recognizes, "They have different challenges I am sure."

For example, Dara, who farms with her husband Kevin, also reports that other farmers in the community think of her husband as the farmer. As she describes, "Regular farm type guys will come in and say, 'Hey Dara, where is the boss?' One day I put my hands on my hips and said, 'Hey Joe, you are looking at her,' and he went, 'Ho, ho, ho. I am going to look for Kevin.' Okay buddy." Women farmers resent having to negotiate the stereotypes in agricultural communities that they are farm wives and their husbands are the farmers. As the above quotes affirm, even though women farmers identify themselves as famers, family members, and community members, other people in the agricultural community often undermine their identification as famers.

Challenges to Identifying as a Farmer: "Hobby Farmers"

Another challenge to being seen as a farmer noted by the women we interviewed is the type of farm that they operate. Most women that we surveyed do not manage large-scale commodity farms but instead have smaller farms, often with horticulture products or livestock, and they are sometimes deemed "hobby farmers" or "lifestyle farmers" by those in their communities and by institutions that serve farmers (O'Donoghue et al. 2009; Hoppe and MacDonald 2013). Others who entered farming later in life were viewed as gardeners or had off-farm jobs to supplement the farm income. Many women farmers were uncertain whether they were not taken seriously as farmers because they were women or because they were farming in unconventional ways.

Lucinda and Bailey operate a farm with draft horses and are often viewed as hobby farmers rather than "real farmers." As Lucinda indignantly explains, "They think that we decided that we are just going to have this expensive hobby and we are going to take everything that we own and just decide that we are going to run around and chase animals and it is our version of playing golf." Lucinda also recognizes that she may not be taken seriously because she did not grow up on a farm. She said, "I think part of the problem is also that if you are not on a family farm that has been there for generations and you come into farming later in life, they are quick to call it a hobby."

However, growing up on a farm does not guarantee that the local community will consider a woman a real farmer. Haley points out that the local farmers, even in the area where she grew up, did not take her seriously. "I know that I was told that when I first started farming in 1980 that at the local feed mill they were taking bets as to how long I would last. The longest bet was six months. . . . So that [growing up in a farming community] is one thing that I feel is a huge obstacle." Her tenacity has earned her respect of local farmers in the community over time. "I have lived on this farm for twenty-five and a half years. So I did not bolt. I do not know how many times I have thought about bolting, but I did not bolt. So that speaks . . . that fact speaks volumes too."

This view of women farmers as hobby or lifestyle farmers extends beyond the agricultural community to those businesses and organizations that provide needed services to farmers such as financial institutions that provide loans, agricultural businesses that sell machinery and inputs, and educational institutions that provide technical assistance. In fact, the USDA has settled several large lawsuits due to patterns of civil rights violations and discrimination against women, Hispanic, Native American, and African American farmers. As reported recently, "Some were told there was no money left for a loan or no more application forms. Other farmers received loans, but for not enough or under terms that forced them into foreclosure. A few women say they were offered loans in exchange for sexual favors."[5]

Several women we interviewed described this kind of bias, thinking of women farmers as only hobby farmers in lending practices. Lucinda and Bailey feel discriminated against in their efforts to obtain loans for their farm: "I do not know if we are discriminated against because we are women or we are discriminated against because we are looking at a different definition of women in farming. I think if we were doing production farming and we were taking our potatoes to the farm market and this was a farm that we had inherited from our parents (so that we did not have to have the big bills of purchasing the farm), we probably would then fit in better."

Haley found that being a woman worked against her ability to obtain financing. "The second thing that was an obstacle is finances. I just felt that there was a bias when it came to me going to the bank and asking for money.

. . . I remember having a conversation with the banker and saying, 'So was I turned down because I was single and female?' He said, 'Haley I cannot say that.' I said, 'You do not have to.'"

Local feed stores and equipment dealerships have long been a bane to women farmers. Dara describes her experiences of being treated with disrespect when buying feed at agricultural organizations and agricultural businesses. There are "just certain places where I do not want to go. The livestock council—it is the same old geezer club. Actually there is a feed mill in Summerburg that I won't go into. They laugh. I go in and I was ordering chicken feed, and you know, okay, so I buy 150 pounds at a time. They carry it and throw it in my car. One bag of corn and a bag of oyster shell, so my bill was thirty dollars. There was a guy there and he came into the store behind me and I was getting my order and stepping up to the counter, and this guy literally stepped up and sort of elbowed me out of the way and said, 'I am more important than you. I spend more money than you do here, and they will wait on me first.' Everybody in there laughed. I do not know if it is because I was only spending a little bit of money or because I was a woman. Who knows? I knew I was never going to go back in there. I will never give those people another dollar of my money."

The Role of Identity in Feminist Agrifood Systems Theory

The complexity of farming women's identities on the farm, in the household on the farm, and in the farm community point to the emerging role of women in agriculture. Although certainly some women maintain dependent roles as farm wives, this characterization of women in agriculture is clearly insufficient in understanding women farmers who increasingly identify as independent voices and decision-makers in farming. Through the lens of FAST, we can see how this shift in women's identity on farms is changing the identity and culture of agriculture in the US. They are expanding the notions of who is considered a farmer and who is making decisions in agriculture.

Women farmers can and do play multiple roles on the farm, some more central to their identities than others. Women often self-identify as farmers and realize that role through the perceived approval, or lack thereof, by relevant social groups, such as family members, other farmers, other women, agricultural service providers, financial lenders, educators, and others. The way that women's work on farms is perceived and labeled by men (and women) reflects the way women farmers are embedded in an agriculture still characterized by patriarchal norms. The social, psychological, organizational, and practical barriers that women face as a result have led them to develop new and different ways of working on the farm that support their identities as farmers.

Many women on farms and farm wives actively confront people who stereotype them and do not recognize the importance of their work. For many women, their multiple roles go beyond farming, and they see their identities as farm partners, mothers, wives, and workers in other off-farm occupations. The innovative ways women have asserted their independent roles in farming have forged new paths and created new spaces in agriculture for those who are not part of the traditional and conventional agricultural system. Women have claimed the role of farmer in the face of structural and ideological barriers that have and continue to try to limit the definition of farmers and farming in the US. In the following chapters we will see how women farmers use their identity as independent agents in farming to negotiate challenges to succeed in farming and to integrate multiple values in agriculture.

Chapter 3

Sowing the Seeds of Change: Innovative Paths to Land, Labor, and Capital

Land, labor, and capital are requisite to farming, and all three are contingent on the others when it comes to agriculture. A farmer needs capital to purchase or rent land and to pay workers, but capital is difficult to acquire without having fertile land and competent labor to meet production demands. Obtaining access to this agricultural trifecta can provide serious challenges for women farmers. The major route of entry into farming comes through family farms that are passed from one generation to the next—usually from father to son. "Farmers are farmers' sons" as a result of gender disparities in inheritance systems (Alsgaard 2013, 347). Patriarchal inheritance systems have long served as the primary means that male farmers use to establish themselves in the business, and, according to Alsgaard (2013), this pattern of excluding women continues today. In this chapter, we discuss a number of the challenges for women farmers in their efforts to secure land, labor, and capital, and we illustrate these challenges using examples of women farmers who have employed innovative strategies as alternatives to traditional farm inheritance, agricultural labor markets, and commercial lending agencies (IIWF). These innovative strategies reveal the diverse motivations of women farmers to obtain the necessary resources to farm. Women farmers draw on the growing interest in agriculture among the nonfarming populace to secure labor, and they take innovative approaches to manual work aided by devising, adapting, and using appropriate tools and equipment. They also draw on emerging services that specifically target women farmers in order to obtain resources to finance land and support for their operations.

"Having My Own Farm Has Made All of the Difference": Gaining Access to Land

In the US, land access follows gendered trajectories whereby mostly males inherit land from family members, which often excludes women from becoming primary landholders through their family. Historically, women from nonfarming backgrounds entered a farming household as partner to a male farmer. Women became farmers through marriage, and some became principal operators only when they were widowed (Sachs 1993). However, outside of this patriarchal system that normalizes the dominance of men over the subordination of women, women who want to farm but who lack the opportunity to inherit land and who are not motivated to seek farm partners must find alternative ways to gain access to land. Here, we provide three examples of women farmers who used different approaches to secure a place to raise animals or grow crops. Each approach represents a distinct form of land tenure (purchase, public, and private leasing) that meets specific agricultural, social, and economic values of innovative women farmers.

Buying Land to Farm: Marilyn Hart

Access to land presented a major challenge to Marilyn Hart when she tried to get into farming. "I was not going to marry a farmer to get land. For years I was farming on other people's land, doing what I could and having big gardens and raising heifers. It got to the point where I did not have enough control over events." She was frustrated because when she was farming on land that she did not own, she lacked control over crucial decisions about the farm infrastructure, such as barns, paddocks, and pastures. Although she worried that she was getting a bit old to start off on a new farm (she was in her fifties), she decided to take the risk. "I thought, okay, it may be late, but if you really want . . . your own farm, you have to do it. Get off the pot and do it. So I started looking into things that I could do myself and started a small farm. Little by little, I built it up. This was only three acres when I started. Now I have an additional twenty-one acres that I bought . . . to run cattle." After working as a ranch hand and abiding by restrictions placed on her farm operations by landowners, Marilyn approached the concept of land ownership as a way to achieve both autonomy and control over her farm enterprise. "Having my own farm has made all of the difference because you can make your own decisions." Having the capacity to make her decisions about the farm enabled Marilyn to try different enterprises without being affected by the restrictions of the landowner. As she explains, "I found it very empowering to be able to have my own actual physical farm."

Owning land not only empowered Marilyn to make decisions that affected her operation, but owning land also changed her social influence in the agricultural community. Farmers who inherit land often have embedded ties to their neighbors and social networks, but this is not the case for new farmers establishing themselves without history on the land or in the community. "When you buy land you become somebody. That blew my mind. First of all, they [the neighboring farmers] think you are serious about being in a community. When you buy land you are making a commitment, and they pay more attention to you when they realize this person is making a commitment to be here. When you own the land, people stand back and say, 'Oh, what are you going to do?' So maybe that is part of it. I am not sure, but I certainly noticed that ninety-nine percent of the time it was male farmers." Buying land in small parcels is one option for women farmers, but it's a long-term strategy. From Marilyn's experience, owning land not only allows her full capacity to direct her operation as a private enterprise; it also provides her with legitimacy as a farming member of the local community.

Leasing Publicly Held Land: "You do not have to have the moon and the stars to get started."

Although Marilyn has many advantages in farming her own land, her advice to new farmers relates to the need to buy land and how much land: "You do not have to buy land to start. In fact, buying land to start is a bad idea because there is so much unused land out there that people really would like to have it used for something good. If you can find a relationship . . . you do not need a lot of acres to start. It depends on what you are doing, but five acres is plenty to do a decent market garden and a bunch of chickens, and you are in. . . . You do not have to have the moon and the stars to get started. Just start with what you have and learn." One innovative strategy for obtaining farmland is to lease publicly owned land. For example, Irene Tilly runs a successful vegetable production, selling and distributing the produce through a Community Supported Agriculture operation (CSA) on land owned by her local township. By gaining access to public land for farming, she found a unique strategy for farming without the burden of owning land.

Irene's interest in farming started when she was a Peace Corps volunteer in Africa, and upon her return, she started working for a large-scale tomato farmer who sold his tomatoes in farmers' markets in New York City. Then, for four years, she managed and operated a vegetable, fruit, and value-added farm, and she sold the produce at farmers' markets.[1] She enjoyed farming and direct marketing, but wanted something more, observing, "I liked going to market, but I did not feel like it was everything that I wanted from my experience. . . . I think the something more is being involved in the community." And the something more for her involved having land that involved the community. She first found a Pennsylvania township that was advertising for a farmer to farm on land that was owned by the township. Although that land did not seem appropriate for her plans, she was drawn to the idea of farming local government-owned land and looked into other possibilities, one being land owned by the township where she was living. "I put in a proposal to the township, and they accepted it [with] a ten-year lease . . . a five-year renewable, [and] a way in my lease for me to get out easier than it is for them to end my lease. It was really set up for me, . . . a farmer."

Irene relied heavily on networking with others to gain access to the township land, gain information, and make a decision about her new direction. She explains that the township agreed to provide her with land because people in the community would be involved on the farm. She explains, "This is how I pitched it to the township: that it would make this CSA more successful if more people in the community were feeling like they were giving to the project or part of the project. The more people that are involved, the more chance that it is going to keep going." By producing vegetables for a CSA operation on public land, she accesses land and also creates relationships in her community. To someone who said to Irene, "You do not sound like a farmer," she said, "Well, maybe there is just a different way of being a farmer these days. Maybe there are just different things that we need to start doing to be successful." Accessing land to break into farming is difficult, but innovative approaches to gain access, such as using public land, may be a key strategy for women without significant capital resources to acquire land without purchasing it.

Securing Land in Urban Spaces: Carol and Sasha

Locating land in urban areas can present other challenges for women farmers. The growth of urban farming has intensified in tandem with interest in local food, health concerns, and economic equality (Bohn and Viljoen 2011). Urban agriculture has recently gained increasing support and interest from groups such as schools, churches, government agencies, urban planners, and public health providers. Anecdotal evidence suggests that women predominate in urban farming.

Carol and Sasha are two such women who farm in the center of the city. As Carol describes the location of their five-acre farm, "We do have noise—" She points to the helicopter flying over during the interview that is so loud we have to wait to continue our conversation. "It is very urban. Because this is a big open piece of land, this is a landmark for helicopters. So we really are very urban." They strongly connect with their community, and Sasha describes that one of the reasons they chose this piece of property was because it was in the community where she raised her kids. "I am very near where I brought my kids up, which was one of my criteria in looking for a place that we would own together. I wanted to stay in the neighborhood where I had been since 1976. . . . So I have all of those connections." These women farm the land they own, but they also farm at another site in the city. "It is a market garden that is built on three vacant lots in a very distressed neighborhood. So we had two sites that had combined and forced us to create a nonprofit to primarily do teaching, . . . modeling, . . . [and] advocacy for urban agriculture and [for] food self-reliance in urban areas." Women farming in urban areas often promote sustainable agriculture and healthy food for their local communities. Access to land for urban farming is problematic in some locations, but in cities that have experienced population losses, some women are taking advantage of land that is available to build sustainable urban food systems.

"If It Were Easy, Everyone Would Do It": Accessing Farm Labor

Farming is labor-intensive and often requires heavy physical labor, sometimes presenting barriers to women. While many farmers reduce their labor needs and costs by investing in equipment-intensive production, other farmers and many women farmers who lack extensive capital resources rely on manual-labor-intensive systems of production. Finding enough workers to do farm labor often proves difficult and costly. For farmers with small acreage, the cost of hiring labor is often prohibitive, so they often work extremely long hours themselves. For women farmers, the issue of labor can be multiplied because they often face physical limitations; personal limitations such as time, strength, age, skill level; and increased labor requirements due to a lack of availability of capital, tools, and other resources. In some cases, women principal operators who farm alone lack access to family labor. For many women on farms, their own time for farming is also constrained because their work is divided between farmwork, off-farm work, and social and reproductive labor (e.g., community building and managing family and household). Given the constraints around labor, women farmers and other small farmers must decide whether to do the labor themselves, hire labor, or rely on interns or apprentices to get the work done.

Farming requires a lot of hard work. As Diane, who farms with her husband, explains, "I think that the biggest obstacle that all of us have as farmers, is a tremendous amount of hard work. You know a typical farmer works twelve- and fifteen-hour days at times." Although she explains that the hard work "has never been an obstacle because Steve and I both are extremely committed to what we do. It is just that sometimes you just get tired. You just need some extra time. So to eliminate that tiredness, we keep Sunday as totally a family day and a God-centered day." Diane describes why they cope with the huge amount of work on the farm. "Where you can work right at our own home and you do not have to commute to work, everything you put into the farm there is an immediate reward. I think that is why a lot of people stay in farming, because it is theirs, and they have that sense of ownership, and that is beautiful, but you have to learn to give yourself some downtime, and you have to learn to incorporate help. If you get so overwhelmed, forget it. So that is a catch-22. That is why some people get totally out of farming."

Linda, who runs a large CSA and value-added operation, underlines the amount of work required of a farm family by suggesting, "If it were easy, everyone would do it. So every day we are faced with any number of challenges, not the least of which is exhaustion. Right now people are working eighty hours, including me, . . . just because there is so much to be done and we are so busy. It is the good news, bad news. We are so busy we can no longer—two people cannot do it anymore. So as you grow, you get some luxuries, but it also creates a whole different set of problems and complications." As Linda suggests, labor is a critical factor in growing a farm operation. As farms expand, farms need more labor than a farmer, and perhaps a partner, can provide, requiring them to hire workers or recruit volunteers or apprentices.

Obtaining Labor

Obtaining labor remains a major concern for almost all farmers, but Marilyn gives examples of why labor may be a more difficult constraint for women farmers who are farming on their own. "Farming is a nice life, but not without help. For women, especially sole operators like myself, labor is an issue. I do not have that husband who can say, 'Oh yes, I will just go out there and fix this thing with my lumber and my carpentry skills.' I cannot do that stuff. I do not know how to do it, and I do not have time to learn. So I have to hire it done." Sasha describes her experience with farm labor by comparing it to the model she was exposed to growing up: "My uncles hired people at dirt wages to pick his crops. That was me in agriculture. I was used to that model for agriculture, but it really did not interest me."

Sasha and Carol recruit volunteers who are invested in success of the farm and its role in dense urban neighborhoods. Says Sasha, "There is so much interest in what we are doing in the city, and people want to be in the CSA. Our goal is to have the CSA consist of people who can either walk or bike here—we can do this—which most farmers can't. There is so much interest in the urban area and getting back to the earth and back to where we have been." Engaging people who are not traditionally thought of as farm labor to help grow food for themselves and other eaters is becoming more common on sustainable farms, many of which are managed by women.

One of the strategies many women farmers and other small farmers use to meet their labor demands is to rely on interns, apprentices, and volunteers to do some of the labor on the farm. Wood (2013) found that many small-scale vegetable farms in the Northeast rely on interns and apprentices and that the majority of their apprentices are women. Diane comments how interns have eased their workload. "Having our interns is a huge help as well as having our student workers. We have other volunteers who come to the farm. So it is just that added help that really takes the edge off of things, and I do not feel that stress." Many farmers rely on interns or apprentices to do labor intensive activities such as planting, weeding, and harvesting. An entire culture has emerged for young people interested in local, sustainable agriculture to spend a summer or two working on organic or small-scale farms.

Variability exists however, in terms of how much interns or apprentices learn and how much they are used for labor. In fact, Wood's (2013) study of farm interns found that many interns expressed frustration that they did not receive the level of training on the farm that they had expected. Marilyn uses apprentices for labor but realizes the limitations of apprentices. She finds that apprentices have an idealized view of farmwork and farm life and often fail to understand the heavy and often tedious labor requirements. As she explains, "They think they are going to go out there and do something two or three times. Okay now I know that, now what? No, do it again. Do it every day. Do it from now until the end of the season. Oh, it is not so exciting anymore." Internships are a temporary source of farm labor, undertaken by people with little to no experience farming. As Marilyn explains, interns perform basic farm tasks and learn by doing these tasks over the course of their characteristically short farm stay.

Apprenticeships and Internships

Informal apprenticeships and internships often last at least a full growing season and commit the farmer to mentorship and management tasks. This duration and commitment can create both rewards and challenges for a farmer who agrees to host interns or apprentices to work and learn on the farm. As Marilyn explains, "I only recently started taking apprentices because taking apprentices is a big responsibility. You are their mentor by default." She works off the farm and finds that after she comes back to the farm, her apprentices are eager to learn while she is faced with "a long day of chores to do on the farm. You have this person who wants to learn everything right now. So they are asking you a million questions, and you have a responsibility to them. They make mistakes, and you have to be patient with them. . . . I mean, how much complication do you want on your farm? You know, they help you, but it is not free, whether you are paying them or not. It is just an additional responsibility and complexity." As Marilyn points out, using internships as a strategy for accessing labor is not always an unequivocally mutual benefit for farmers or interns.

On the other hand, some farmers are particularly compelled to work with interns. Diane and Steve define themselves as farm educators—committed to teaching young people how to be land stewards by growing food sustainably. Diane describes how their internship program has evolved over time. "At the end of this month we have a young woman coming in from a local college. She will be here for five weeks to intern. She will be receiving college credit, which is kind of unique. Most of our interns do not fall into that. Almost weekly we have people that call us or e-mail us to intern here. Yes, it has grown, so we are pleased to have that extra help. We need to take the time and mentor and give to the young people. It is a nice exchange because we want to teach them. We want people to replicate what we are doing here and then have extra hands." Marilyn and Diane both point to unique relational characteristics of farm labor practices that rely on interns. Farmwork is hard work, and it is also work that must be learned. Interns and apprentices become integral people on a farm for the labor they contribute. At the same time, farmers who offer their skills and knowledge are teaching a new generation of people—often young women who did not grow up on farms—how to grow their own food.

Some farmers rely heavily on young people for labor and inspiration as well. As Carol describes: "Basically a lot of what we have done has been inspired by the interest and engagement of young people. All of the connections that we have made in terms of the nonprofit activities and the community education stuff and everything. The interns we have had. We have had so much engagement by people in the city, so excited that there is a farm here, and twenty-something people. One year we had three people who volunteered to be interns and came up here and worked every day for nothing." Recruiting interns in the city is quite easy. Carol accounts for their success at recruiting interns as part of the urban and local food movement: "So being urban and having a place for people to live in the city. I mean you know the interns who come here can go down to the bottom of the hill and be at the hippest bar with a cutting edge band." Marilyn agrees that the benefit of interns and apprentices goes beyond their work contributions on the farm: "They inspire me. They remind me of me when I was young, and I am glad to give them an opportunity. . . . So you click with different people different ways. You know I was lucky because the very first apprentice that I had become an employee. She was that good. I was so pleased with how she was. At the end of the day she was a tremendous asset, and you know, we formed a long-term relationship, and she was inspired to start farming, and she is trying her first farming experience now. I feel good about that. So you know it is give and take."

Labor Requirements

Farmers who sell produce directly to consumers at farmers' markets or through CSAs have realized that the labor requirements are often overwhelming. In addition to planting, weeding, and harvesting crops, at least several days a week must be devoted to selling crops to consumers. As farms diversify into different types of markets, some farmers have made the decision to hire other people as farmers to operate their CSAs. For example, Diane and her husband started a small CSA with four families that they increased to sixteen families, but then recognized that the labor requirements were more than they could manage themselves. As she describes, "So we really like the concept, but it is a lot of labor intensity and with our education and with our value-added and with our own children's education. I have children here at home on the farm. We just can't keep up with the labor, but with someone like Megan, she is so excited because she loves to grow food. So we can compensate her to a minimal amount. We still provide her free room and board, and she will be moving into the straw bale addition next month. That is perfect. She will have a neat little place to live."

Similarly, Linda, who has a large CSA with over two hundred members, hires a farmer to run the CSA. Some women enter farming in middle age, and the combination of being female and older can make labor difficult. As Sasha says, "Hello! I am almost sixty years old. I am doing that. You know we are going to put up a damn fence or a structure for peas and tomatoes and things that need support. We are going to change how we do it, and I think it is a challenge to keep doing it. I think people give up all of the time because they can't do it because of age. We are fortunate that we have small scale here. We have a lot of help and a lot of interest in providing help. We have no trouble attracting interns, and so we have the things that we need to do this into our later years. We have family engagement. My kids are interested in it and we have to find a pathway for them to really form and engage. I think aging is a big stumbling block for me." As the data above reflect, the labor needs on farms can be significant, even on small-scale operations. Women farmers often use interns as a strategy for incorporating more labor resources on the farm, which can result in new challenges and opportunities for them because they must shift into managing workers, delegating tasks, and teaching skills to interns. At the same time, interns can help women farmers stay in business and offer a pathway for some interns, many of whom are women, to break into farming.

"Farming Is Always a Risk": Accessing Capital

For many women farmers, as for many small or new farmers, access to capital is problematic and often sets the parameters for the type of operation that is feasible. As mentioned in chapter 1, women seldom farm larger scale commodity operations due, in part, to their limited access to capital as well as land. Marilyn, drawn to farming primarily by her interest in dairy cows, lacked the capital to start a dairy operation: "Putting in a dairy is a pretty big financial commitment. You cannot do it without some debt, and you got to bank on it paying. And then you have to say, 'How committed am I to this?' because it has a time line to it. Farming is always a risk. There is a risk associated with it, and the farmer bears one hundred percent of the risk." Her lack of capital and the risks involved in borrowing capital set the parameters for the type of livestock and the size of the farm she operates.

Marilyn obtained a loan for expanding her operation from the Farm Service Agency (FSA). The FSA is part of the US Department of Agriculture and targets a specific portion of its loans to what it refers to as "socially disadvantaged farmers," which includes African Americans, Hispanics, Native Americans, and women. As Marilyn explains, "When I got the twenty-one acres, that let me have the cattle—that was a USDA loan, a FSA loan. They have a special program for women and beginning farmers and minority, disadvantaged.... They have a special pile of money for those kinds of loans.... Women are at an advantage when they apply for a loan that way, and I was lucky because I have a really good loan officer.... She is good. She is a woman for one thing. She is good with everybody.... I have seen her operate with ... Amish people, too. She figures out how to make it work."

FSA targets loans for populations who do not qualify for commercial loans. Based on the high risk of such enterprises, commercial lenders are largely reluctant to make loans to farm start-ups and to people who traditionally have not been farmers. For women farmers, FSA provides a lending source with interest levels historically much lower than commercial lenders. FSA offers different loan types according to specific needs on farms that include operating, ownership, and emergency loans, for example. Banks traditionally show no interest in owning land or farm enterprises if a lender defaults on a loan. The FSA, however, is a targeted governmental program to support agriculture and farmers in the US and thus fills a gap for people who need capital but cannot get it from commercial institutions that largely deny farmers access to capital. As a result of the class-action suits by African American, Hispanic, and women farmers against the USDA, more capital and credit may be available to women farmers in the future.

Community Supported Agriculture: Another Approach for Accessing Capital

Community Supported Agriculture (CSA) is a direct marketing model becoming more widely adopted by farmers wanting to sell directly to customers. Farms sell shares in the spring, asking for an investment by the consumer-subscriber on a predetermined number of weeks during which the farmer will provide the farm's products to subscribers, according to their investment level. The model is designed to provide farmers capital in the spring, during a time of the year when they need it most and when it is most difficult to acquire. The model attempts to reduce the risk farmers take solely on their own if a crop fails. The investment made by subscribers to receive whatever the farm produces that year helps farmers avoid taking out loans for early season input costs.

Raised in a traditional farming family, Sasha contrasts their small urban farm with her brothers' and uncle's farming operations. "I am from [the Deep South] and my half-brothers grow soybeans, you know with huge machinery, and before that they grew cotton. You know when I called my uncle and told him I had a farm, he said, 'How big is it?' I said, 'Five acres.' He was laughing so hard, and I am sitting there thinking, the last time I was at your farm, you had a block-long shed, a two-story-high shed full of reapers and sprayers and all of the equipment, and you are so far in debt with your two hundred acres and growing in commercial agriculture. Using poison. You can laugh at me if you want to, but this is where we are sustainable. None of his kids want to farm. You ride a machine all day, and you are in toxins all day. There is not feedback like you get from a CSA or a farmer's market. There are no connections to the cows that eat your soybeans." Sasha responds to her uncle's disdain for the size of her operation, but she points to an important social distinction in the income stream of farmers growing for the commercial market. Here the CSA model provides a different way for a small-scale farmer to acquire capital, to mitigate the financial risks of farming, and to be in contact with consumers. The CSA model is illuminated further in chapter 3.

Resources Gained, Resources Lost

The women we have studied encounter difficulties in retaining farms even after being successful for a period of time. In our interaction with them, we observed that a few have faced serious illness, leading to the elimination of resources to sustain the farm. Others have divorced, leading to the breakup of farm partnerships. Small farms tend to go out of business at a higher rate anyway, but the farms women have may be at greater peril for the loss of access to resources to sustain a farm, a critical topic to study over time.

Women Farmers as Innovators

The multiple challenges of gaining access to land, labor, and capital intersect as women endeavor to farm successfully and to gain legitimacy in the farming community. Nevertheless, many women farmers with limited resources creatively develop innovative strategies to meet these challenges. Limited capital is often tied to limited access to land, so women farmers often design and adjust their operations to fit their available capital. Women find new ways to access land either by securing loans geared specifically to women and minority farmers, by farming public land, or by engaging in urban agriculture. Women farmers often meet labor requirements by relying on alternative types of labor in the form of apprentices, interns, students, and volunteers, although the farmers must often cope with the unevenness of the skill level of these alternative types of labor and must balance mentoring with labor demands. Yet at the same time, interns and apprentices also bring energy and hope for the future of farming.

These ways of accessing the resources needed to farm clearly reflect women's contributions as independent agents in agriculture. Although women are still constrained by patriarchal control of land, labor, and capital, successful women farmers have been able to bypass these systems by forging innovative paths into farming. In doing so, women farmers contribute new models of farming by using resources to farm in different ways and claim space for women in a male-dominated industry. These new roles for women farmers have important implications for equality and social justice in agriculture, which make them fundamental components of society. In the next chapter we discuss further implications of how women are using these alternative paths to farming as opportunities to incorporate multiple values into agriculture.

Chapter 4

Reaping a New Harvest: Women Farmers Redefining Agriculture, Community, and Sustainability

Women farmers have responded to the barriers to farming described in the previous chapter—including limited access to the land, labor, and capital they need to farm—with innovative strategies that emphasize smaller farm scales, diversified high-value and value-added products and enterprises, unique marketing strategies, and sustainable practices. For example, many women farmers produce a diverse array of crops and livestock to provide multiple opportunities for revenue. Many women farmers are using creative (and often multiple) marketing strategies, including direct marketing approaches such as community-supported agriculture, farmers' markets, on-farm markets, collaborative marketing, and online marketing.

These innovations are creative responses that help women move beyond the barriers in traditional farming. Our research indicates that these innovations are not just business decisions, but were chosen and adapted so that they also fulfilled women farmers' personal values. These values include living a life that satisfies them personally, supports their families financially, provides healthy food to their customers, and remains in concert with their land, their ecosystems, and their local communities. In so doing, women farmers are establishing a foundation for an alternative food system.

In this chapter, we first describe the multiple goals women have for their farms, as reported in both interviews of women farmers in Pennsylvania and our survey of women farmers in the northeastern US. We describe how these goals reflect personal commitments that are oriented toward both individual and local values but also community and global concerns. Then, drawing primarily on interview data, we describe several innovations women are employing on their farms that satisfy these goals and begin to create a new food system one farm at a time. These innovations include new products, innovative enterprises and business models, and farms that emphasize sustainability in multiple dimensions.

Women Farmers' Goals

In the interviews we conducted, women farmers in Pennsylvania talked about the ways that family, spirituality, love of the land, and environmental consciousness led them to farming in the first place. Because women often do not receive the farm through hereditary succession, women farmers tend to come to farming through various paths (chapter 2). For some who grew up in farming families, they confront barriers such as those described in chapter 3 as well as challenges in more traditional agricultural systems that confront the other values in their lives. For example, Haley grew up on a dairy farm. She left and went to college expecting to leave farming for good but came back to the farm instead.

> So I came home, and my dad put me right to work. You know, I can remember going out again and feeding the calves and taking wagons back and forth. They were filling the silo. Thinking what the heck do you want to go to a big city for? I mean, this is you! So this is like the summer of 1980, and I would come over here on my time off, and I would just kind of walk around and dream a little bit. I kept thinking, there is something compelling. There was something that was calling. It took years to actually put it all together.

Haley struggled with her love of farming and her other interests beyond the farm, but in the end felt compelled to come back to farming because it had become part of who she was fundamentally. Haley did decide to stay on the farm but, after an initial attempt at running a traditional dairy farm, decided she needed to do something different, something that reflected her interests in building relationships with community members, especially girls and young women. She developed a farm camp for girls, where the girls are responsible for the farm animals and crops while there. They gain valuable self-confidence and skills, but more importantly, they develop relationships with other campers and the land. Although the farm camp for girls is Haley's primary source of income, she also has welcomed artists, girl scouts, and church groups to her farm.

> I have the artist workshops. There is one in May that lasts for a week. There will be artists that come in for the day and set up and paint, and I have one artist, an older woman. . . . She is so enamored with animal physiology. So she just comes and sketches the goats. Just so she can get how they are muscular. . . . [The artists] like to be able to set up in a space where they know that they are not going to be asked to leave. I also have Girl Scout troops that come in for weekends. Church groups that come in for weekends. I have had meetings—you know, a variety of different groups use the studio space upstairs for meetings.

As Haley describes her business, she explains, "So I guess what I am selling, or how I make my money, is I am selling a notion of sorts. I am selling camaraderie." These experiences reflect Haley's desire to instill a set of values in multiple groups, but especially young women. They emphasize community as well as a connection to and respect for the land and the food system.

Diane's experience provides another example of the multiple values that shape women's path to farming. Although she grew up in Pennsylvania, Diane and her husband, Steve, had moved to Seattle, Washington, where Diane was working on a graduate degree in education. They decided to move back to Pennsylvania to farm, but they wanted to operate both a commercial farm and an educational facility:

> Steve, my husband, . . . has an agricultural background and sort of chemical farm background. I was raised on an organic hobby farm. So having that background and his conventionally based background, we combined to have a noncertified organic farm, which was our goal when we married eleven years ago. We use the land to teach others about where their food comes from.

Diane and Steve gained inspiration about how they would operate their farm in Pennsylvania after visiting an herb farm near Seattle, which Diane describes as

> a wonderful model. . . . They had a little culinary garden where they then brought people in to eat. It was like one hundred and fifty dollars a plate. So the whole key was to teach people how to use herbs in a culinary fashion and then entertain them and educate them as well.

Diane also wanted to keep family and spirituality at the center of planning her operation:

> We had a big butcher block piece of paper, and the key was family, and we are very God-centered between our relationships. Where God was at the center, and then we had little spokes coming out of that and how we would put the farm into action. We arrived here in June of 1996, Father's Day weekend, and we celebrated with my father and my sisters. So we had adjoining properties. Two of my immediate family members live in the immediate area.

Diane and Steve have built their operation into a very successful for-profit herb farm and nonprofit educational center with the aim of bringing consumers into closer connection with their food, the land, and their community. They creatively integrate their educational facility with their herb business and use the herbs for multiple purposes:

> We grow herbs for use in value-added products. We sell herb plants, and we have educational classes about using herbs in every capacity. So if it is a wild, edible herb plant, we take people out on hikes through the woods and the meadows where we collect and identify these herbs. Then we come back and have a wild edible feast, or if it is grown, dried and processing herbs for an herbal soapmaking class for soap that we sell. We use herbs in our direct markets, retail, wholesale, or give it out in our CSA share. So it is kind of like from the ground up. We grow the herbs and then use them in many, many facets on the farm.

Even though Diane knew what she wanted the farm to be, it wasn't until her sister asked what the farm's main goal was that she was able to put it into words:

> I said our main goal is to connect people to the land and help them appreciate that. If it is through natural hygiene products or if it is salsa class where they are going to be growing the herbs and vegetables, if it is a school group that learns about how important the watershed is and how to protect it, you know that is where we are at.

Like some other women farmers with whom we spoke, profit is only one among several goals for Diane. She and her husband cannot operate their farm without a profit, but connecting people to the land and teaching is as important as profitability.

Haley's and Diane's stories reflect the role of personal commitments and values that led to their choice to start their own farms. They chose their farm's products and structure so that they could enact their values through their livelihood and begin to bring their customers into a new relationship with the farm and with a food system that provides an alternative that is more local, sustainable, and built on personal relationships.

Haley's and Diane's stories are also instructive and offer a qualitative lens to view the ways in which women are forging their own paths through farming. We gauged the prevalence of these multiple values using our survey of women farmers in the northeastern states. The survey provided a list of potential outcomes that farmers might use to measure the success of their farm operations. Women farmers were asked to describe the level of importance (from "not at all important" to "very important") they place on each potential outcome for their own definitions of success. Figure 3 illustrates the percentages of respondents who indicated the particular component was "very important" to their definition of success.

Overall, the pattern of responses to these components suggests that women farmers view their farms as fulfilling multiple goals, as all but one potential outcome had more than half of the women farmers rate it as "very important." The responses suggest that personal satisfaction and fulfillment, good relationships with customers and community members, profitability, and environmental quality are all very important to women farmers. Four components of a definition of farm success were selected by three-quarters or more of respondents: "enjoying what I'm doing" (85%), "providing healthy food" (85%), "satisfying customers" (78%), and "maintaining a high quality of life" (75%). These responses suggest that the top priorities for women farmers who responded to the survey include personal satisfaction and creating products that provide customers with both healthy food and a good experience with the farm and its products. The next three components of a definition of farm success have very similar percentages of respondents identifying them as "very important": "increasing profitability of the farm" (65%), "improving environmental quality" (64%), and "improving soil quality" (63%). This pattern suggests that business success is similar in importance as environmental quality. The next grouping emphasizes the importance of relationships within the community, with about equal percentages rating as "very important": "maintaining good relationships with other farmers" (58%) and "being respected in my community" (57%). Half of the women farmers surveyed indicated that traditional measures of farm success ("keeping the farm in the family" and "increasing production yields") were very important to them. Finally, only 8% indicated that "increasing the size of the farm" was very important to them, a significant concept and notable finding given the smaller sizes of women's farms.

These data suggest that women farmers are trying to balance multiple goals for their farms through their farm operations, including personal satisfaction and fulfillment, good relationships with customers and community

Defining Farm Success

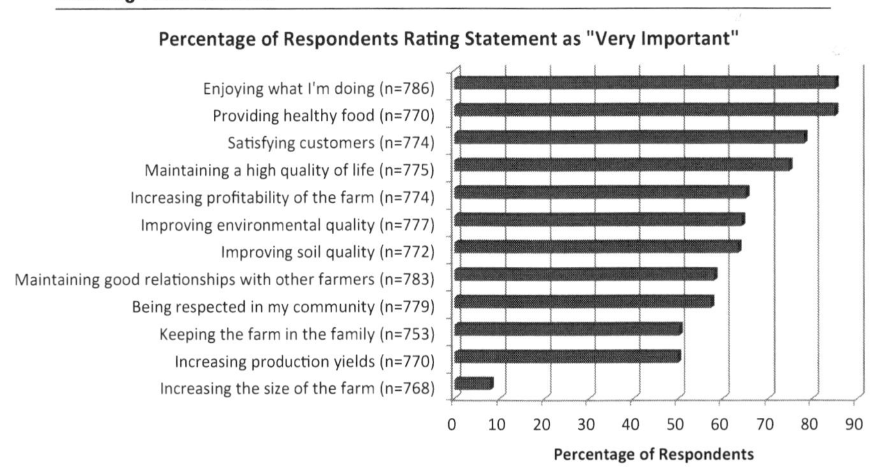

Figure 3. Importance of dimensions of farm success according to women farmers in the Northeast (data source: SNEWF).

members, profitability, and environmental quality. Of lesser importance to survey respondents are goals related to farm productivity, creating a farm family legacy, and increasing the scale of the farm, all of which are goals more frequently associated with conventional farms and food systems.

As indicated by Diane's and Haley's stories, these multiple goals are reflected in the types of farms and farm products women farmers have created. Their goals include creating a profitable business that supports their families and keeps them on the land. One way of doing this is to diversify the farm's products and create and employ alternative business models, models which often blend profitability with building community. Women also seek to share the benefits of farming with others. This sharing happens in myriad ways and occurs both between farmers and consumers and among farmers themselves. In the following section we describe the innovations women farmers have used to create additional streams of income to keep the farm business profitable, build relationships with their customers and communities, create personal satisfaction and fulfill spiritual needs, and maintain a healthy natural environment both on and off the farm.

Diversification of the Farm: "Rude" Vegetables, Unique Meats, Value-Added Products, and Multiple Revenue Streams

One strategy women farmers have used is to grow a wide variety of high-value crops to satisfy customers and develop a niche within their farming markets. Carol says, "Because we are small, we grow different weird things. . . . We can't compete with the farmers that are bigger than us. So we grow rude [i.e., unique] things. We had four different kinds of sweet potatoes. We had red, white, blue, and yellow potatoes. We had twenty-two varieties of heirloom tomatoes." Nan and Ken Dressler, who have a larger vegetable operation, also decided to sell a wide variety of crops: "We wanted to have as big as a variety of vegetables as we could. We did not have a lot of experience with different vegetables, but tomatoes were one of the big ones from the beginning, and they are still one of our big crops now." They now grow more than forty different certified organic crops, including vegetables, fruits, and herbs, and sell those products in farmers' markets in Washington, DC.

Another strategy is to develop integrated crop and livestock systems, which provides fertilizer for crops and income sources that are more diverse and extend throughout the season. Kelly Novak describes, "[We] bought the sheep to diversify our market and our farm. Not only are they fertilizing and mowing our grass for us, but hopefully they are a source of income for us in the future." Robin and Kent Green began farming by raising pumpkins for eleven years, and they now have two high tunnels,[1] where they raise peppers, tomatoes, cucumbers, and onions. They also raise a variety of vegetables, fruits, and mums as well as dairy heifers, layer hens, and beef. They have added a new farm store on their property where they have a commercial kitchen and bakery so they can produce fresh baked goods as well as their own sausage and cuts of meat. They will also be selling products from other local growers at their market.

Dara and Kevin Young have diversified to include not only vegetables, but also a variety of livestock products that provides an income stream over the year. Their most profitable livestock enterprise is pasture-raised chickens.

> We have done chicken like eight or ten years now. We process them here. We did four hundred this year. We sell them for $2.50 a pound. That is a lot of money . . . it spreads your money out over the year. If you get CSA money in February until June mostly, then you are running your CSA all summer, but it really is nice to have a cash flow if you can have some meats to sell over the summer. In July we do pigs and chickens, and then in August we did chickens. Most years we have turkeys in November. It just spreads out your income.

Kathleen and Will Swift transitioned a historic 160 acres, which had been farmed conventionally by Kathleen's family, to a growing system based in organic practices and mob grazing.[2] Their farm practices include hand-milking Jersey cows, as well as sheep and goats, and raising Hereford beef cows and heritage breed pigs on pasture. They also have free-ranging chickens, geese, and turkeys, in addition to both vegetables and row crops, including hay, straw, corn, oats, barley, buckwheat, and heirloom wheat. They offer these products through multiple outlets, including CSAs, an on-farm market, and three farmers' markets.

Many women farmers have found success by processing their raw products into finished food products and then marketing those products through multiple venues. Several of the women farmers we interviewed raise multiple

Barb Connell and Mena Hautau participating in an experiential cheesemaking workshop at Birchrun Hills Farm. Photograph by Ann Stone, PA-WAgN.

types of animals, process the meat, and sell directly to consumers. Marilyn and Ruth raise chickens, turkeys, pigs, and various other types of livestock. They work closely with a small-scale animal processor and sell various cuts of pasture-raised meat directly to consumers. Many of the small-scale animal producers find that their customers appreciate knowing where their meat comes from and are willing to pay substantially more for their products. Freida raises goats for both meat and milk. She produces goat milk and has put in a certified kitchen to make goat cheese at her farm. She markets the goat cheese at a farmers' market in New York City. In addition, she has recognized that many ethnic groups are interested in goat meat for holiday gatherings. She has studied the timing, types of processing, and specific needs of different ethnic markets for goat meat for Hispanics, Muslims, and others.

Small fruit, vegetable, and herb farmers also find that processing their products helps them increase their incomes by providing new products (for which higher prices often can be obtained) but also by effectively extending the season compared with unprocessed produce. Irene Tilly raises garlic and has developed a specialty garlic vinegar she sells (at a substantial profit) at garlic festivals. Diane and Steve process the herbs they grow into teas, salves, and soaps. These value-added products are easier to store and transport, and so they extend the amount of time farmers can offer products to customers. The production of value-added products can also use farm seconds that otherwise might be wasted, thus enhancing the farm's productivity and profitability.

Some women build from a single successful processed product to create multiple, direct-market revenue streams. Linda and her husband, Eli, ran a fledgling pick-your-own-business that grew asparagus, raspberries, and Christmas trees. They decided to focus the business on organic and sustainable production and began to experiment with processing their products. They developed a raspberry shrub (a mixture of raspberries, vinegar, and sugar) from a colonial-period recipe that, when mixed with water, makes a refreshing drink. Just as the business was beginning to take off, Eli died of cancer at a very young age. Since Eli's death, Linda has built the food processing business to include over fifty different products (jams, jellies, chutneys, shrubs, salad dressing), and the operation overall includes a retail farm store, a one-hundred-plus customer CSA operation, a wholesale business, and a greenhouse/nursery business. The CSA itself brings over a hundred families to the farm regularly to pick up their shares and offers an important business opportunity:

> Bringing all those people to the farm on a weekly basis also increased the traffic and the market for more local foods. If people were coming to get their fruits and vegetables, they would also want bread and yogurt and eggs and cheese and that sort of thing. So we began searching and creating more and carrying more vendors. Then a couple of years later we added meat people who were local regional meat producers, and we just invited them to come and sell to customers as a service to our customers to allow them to be able to do one-stop shopping.

They now sell products from over fifteen other local farmers and food processors, both in their retail store and online. Through this process of building the diversified farm businesses, Linda has developed a reputation for business savvy and a collaborative approach to working with other local farmers, artisans, and businesses.

Dara and her husband Kevin have also drawn on multiple revenue streams for their farm business. They initially sold their products at several farmers' markets.

> We did farmers' markets in Boalsburg, Altoona, Huntingdon, Hollidaysburg, and another one in Altoona. People were always driving in our driveway saying, do you have this or do you have that? I would say, no, but you can drive to Huntingdon tomorrow and get it. No I am going to Boalsburg. . . . Can't you just sell me some of that here. So then I started putting stuff in the refrigerator . . . and we just decided that we had enough demand to do it here. There is nothing worse than going to Boalsburg and selling ninety dollars' worth of stuff on a day when you can do four hundred dollars here. . . . [Now] we do CSA here and we sell everything from the farm. We do not go anywhere. . . . . CSA is by far the biggest share of our money. . . . We do chicken and eggs, and we have cheese in the refrigerator all of the time.

They have also added a new farm building that includes a retail store and a commercial kitchen as well as a large open room that can be used for specialty dinners and meetings. The commercial kitchen has enabled them to try a number of diverse products and strategies. Like Linda, they make their product decisions based on what their customers want. "We are testing value-added. We know what our customers want. They want a bakery, and they want our jams and jellies, and they want chicken, and they want every kind of meat there ever was, eggs, milk and cheese. They want everything." They bake bread for their CSA customers and also make jams and granola. They have refrigerators and freezers in their building stocked with their own chicken, eggs, and meat as well as products from nearby farms such as milk, lamb, and sausage. Because of what they have to offer at their farm market, their farm has become "a destination market":

> People have to plan to come here. They do not just drive by accidentally. I get a few drive-ins. There is a big sign down on the road because it says it is open Saturdays. . . . Mostly I have regulars who come and spend a lot of money. Like I might have a market day seriously with twenty customers and break $350 or $400. They come and they spend sixty or seventy dollars, maybe fifty dollars, and buy a couple of chickens.

Debra Brubaker of Village Acres Farm describing their CSA planting schedule. Photograph by Ann Stone, PA-WAgN.

Women farmers we interviewed often have diversified products they sell through multiple (mostly direct-to-consumer) outlets. Data from the survey of women farmers in the Northeast support the interview data, with 72% of farm women who responded reporting use of direct retail outlets, which would include on-farm markets, farmers' markets, and direct sales to consumers through websites, festivals, and other events; 17% report the specific use of CSAs. In addition, 87% of women report using at least one kind of value-added strategy. Of these women, 36% report using only one value-added strategy, 25% report using two strategies, and 40% report using three or more strategies. These include on-farm processing, farm tours or agri-tainment, bed and breakfasts, on-farm education, organic production, specialty production, or local labeling. These outlets provide multiple revenue streams to sustain the farm and increase farm income, as well as create opportunities for building relationships with consumers.

Business Models That Build Community with Consumers and Grow Consumer Demand for Local Foods

Women farmers we interviewed reported using multiple kinds of business models, such as cooperatives, collaborative marketing, and nonprofit businesses. For example, Nan and Ken Dressler helped to start an organic vegetable and fruit growers' cooperative which assists numerous other farmers who market their products in urban areas and supermarkets. The cooperative has developed both formal rules and informal expectations that encourage cooperation and minimize direct competition. The cooperative now works with over forty-five producer-members, and in 2012 they produced over one hundred thousand cases of produce for urban markets. Another example is Natalie and Marty Ingram, who started as the only organic grain farm in the region. They slowly built relationships with organic dairy farmers (their primary customers) and with other grain farmers in the region to develop a regional "cluster" of interconnected farms. Over time, they found that an important step—milling of the grain—was needed locally, so they worked together to develop an organic grain mill that processes local grain to produce feed to support the certification requirements for local organic dairies.

These business models all share a cooperative method that emphasizes peer-to-peer learning and transparency of financial and other business information. For many women, profit is important in that they desire to have a farm business that provides a sufficient living for them and their families. Tanya Nash's goal for her farm is to each year pay one more month's mortgage so that eventually the farm will pay for itself and she will not need her off-farm job. Dara and her husband Kevin, who both had other jobs when they began farming, describe their efforts: "[We] decided to try to make our farming income equal our off-farming incomes, and then last year the farm made more money than our jobs off the farm." Like Dara, many other women farmers talk about the importance of sharing financial details about farming through networks of women farmers in an effort to help guide new and beginning farmers through the economic realities of farming.

However, the pursuit of profit is not seen as a singular goal, but one among many goals that need to be in balance. Profit comes not just through direct competition with other farmers, but through cooperation by sharing knowledge, building cooperative business models, and growing the consumer base. Farmers are also committed to educating consumers about sustainable agriculture. In the interviews, women farmers pride themselves on their openness in sharing their knowledge and innovative business practices with others. Diane recalls what other people have to say about her and Steve's innovations and how they respond:

> Wow, you are educating kids, and you have school buses coming in here. You have this little gift shop at the farm. . . . I think that true novelty is that it is so sustainable. I think it is the information. So we use the quote all the time, "There are no secrets at [our farm]." So if anyone walks on the grounds and wants to know how to make soap or wants the recipe for this herbal salve or how do you start a farm or how do you establish a 501(c)(3). I think that is what we want to be known for. Just to be very open and sharing and provide information that people need from what our expertise is. You know we a have a limited amount of expertise, but it is totally open to anyone who wants to pick our brain.

In fact, several of the other farmers we interviewed mentioned Diane and Steve as role models and mentors. They have given a number of presentations at sustainable agriculture events about how to start nonprofits on farms and regularly offer step-by-step directions for other farmers who want to move in this direction.

Many farmers are pursuing strategies such as on-farm education to bring people to their farms to increase their knowledge about farming and food systems. Farmers engage in these activities for multiple purposes including diversifying their income streams, building loyal customer bases, and educating eaters about healthy food and sustainable agriculture. For some farms, educational programs are at the center of the operation and provide much of the farm income. Haley's farm, described in the beginning of this chapter, is an example of a farm where farm camps are at the center of the farm business model. Another farmer who runs a camp on her farm is Leisha. However, unlike Haley, she came to the farm camp through marriage. Leisha moved to the farm with her husband, who recently passed away. Her husband's parents started the camp in 1939 as a place for children to come who needed a place to go during World War II. The farm camp has been running ever since, with two- to three-week sessions for boys and girls aged seven to twelve. The farm caters to children largely from urban and suburban areas with little farm background and emphasizes their relationships with nature and with animals. As on Haley's farm, each child chooses an animal for which they care while they are at camp. There are a wide variety of animals, including goats, donkeys, calves, pigs, chickens, and ducks. Leisha describes with pleasure that the kids leave the farm camp with a reverence for animals, people, land, and water.

Lori Noble, with her husband, brings disadvantaged children to their farm to learn about agriculture. Lori's husband runs the youth program at a local community center, and they decided to bring the youth to their farm in the summer. Lori teamed up with the Cooperative Extension Service and 4-H to develop a farm-based program on family nutrition for kids. About twenty children participate in a wide variety of activities focused on exercise, nutrition, and organic gardening. The activities are designed to be fun, engaging, and educational. She sees the "farm as a resource for the community, both in terms of providing support for people, but also as a resource for people to utilize as a way to improve their lives and improve community."

On other farms, women see the educational programs as a way to add value to their other products and educate the community about farming, nutrition, and the environment. Diane and Steve, who operate the herb farm mentioned above, combine for-profit herb production with a nonprofit educational center. They began their farm with the herb production business. They were committed to bringing people to their farm for educational purposes but were experiencing some difficulties doing this as a for-profit enterprise. Diane's previous experience as a teacher helped them negotiate strategies to integrate their farm operation and education activities by developing their nonprofit center. As Diane explains, "As a certified educator in the state of Pennsylvania I knew that academia, school systems . . . are a little skeptical . . . of a profit-making organization. . . . I knew that to open up doors and work with administrators and teachers to get the kids here [we would need a nonprofit enterprise]." Another advantage to forming a nonprofit was the opportunity to receive funding and grants. According to Diane, "[The] opportunity to receive funding was instrumental in becoming a nonprofit. There are so many grant opportunities for environmental education, which we fall into completely." The combination of the for-profit enterprise combined with the nonprofit educational center—and the unique funding and community relationships the center creates—allows Diane and Steve to balance multiple goals of supporting their family, building community, and connecting community members to the land.

The business models women farmers describe consider the employment status and working conditions of employees in addition to incorporating consumers. For Linda, the employees on the farm are a valuable resource, and one of the reasons she emphasizes multiple income streams is to keep people employed throughout the year. "So I think part of our innovation has been driven by wanting to keep good people year-round and understanding the necessity of cash flow throughout the year and seeing an enormous opportunity with the people that work here." The farm has relied on the talents and passions of the people who work on the farm to help it grow and thrive. Linda's approach to managing the farm and the people who work there is holistic and empowering, as she has developed an organizational structure that brings out the talents of the farm's employees. "I . . . ask what sort of powers that are bigger than me to pay attention to what is next. I listen for that. I try to stay tuned. I think . . . the biggest gift I have in the company . . . is that I listen. . . . I listen for what . . . my staff says." Linda's innovations on the farm go beyond the products they grow and sell to include farm business models, market outlets, and organizational structures. These integrated efforts fulfill multiple goals, including strong relationships with customers and community members, developing talents of employees and future farmers, and a thriving and profitable farm business.

"The Land Is So Precious": Sustainable and Organic Farming by Women Farmers

Sustainable and organic agriculture is growing as a portion of the US farm sector, particularly among women farmers. As we detailed in chapter 1, sustainable and organic are often thought of as interchangeable, but they have distinct meanings. Both are informed by a desire to reduce the negative impacts of pesticides and other inputs and practices common in modern agriculture, and they support building biodiversity and soil health to improve productivity and produce healthy plants and animals, including humans.[3] In addition, women are leading the way in both organic and sustainable production. The US Census of Agriculture reported that women were the primary operators on 22% of the 20,437 organic farms in 2007, compared with the 14% of women who are primary operators on all farms.

In our survey of women farmers in the Northeast, 77.4% of women farmers with row or horticulture crops and 51.1% of farmers with livestock used organic production practices. In addition to using organic production methods, women farmers also reported using a number of conservation practices and structures (Barbercheck et al. 2012). On crop farms, more than 75% of the women used conservation practices such as crop rotation, soil testing, cover crops, permanent vegetation on slopes, conservation buffers, composting, and organic production. On farms with livestock, more than 50% of the women used permanent vegetation on slopes, rotational grazing, pasture planting, nutrient management plans, grassed waterways, riparian buffers, and organic livestock production. Women farmers add biodiversity to their farms by producing diverse crops and livestock or by operating integrated crop/livestock farms.

Almost all of the farmers that we interviewed consider their farming operations to be sustainable. Most emphasized practices that protect and

Mary Barbercheck demonstrating healthy soils. Photograph by PA-WAgN staff.

enhance soil quality and health. A typical comment comes from Katie Inwood: "I definitely try to pay a lot of attention to the soil. We grow cover crops and put cover crops in whenever there is not a cash crop growing and then we till them in as kind of a green manure. We use compost for fertility." Lucinda explains, "We have respect for the land so that what you take out, you put back. Our soil that we are working with this year is stronger and has more organic matter and is healthier than the year before. We practice good rotation of our fields. We understand the appropriate way of processing manure so that when the manure comes back in to our fields that it is composted and then goes back on at the right time. So we are always strengthening the soil and the roots from where we come." Lucinda started her farm with a commitment to environmental sustainability and explains, "My first ideas about it had to do with environmental sustainability. Basically not depleting the soils. Creating a healthy soil and replenishing the soil rather than just growing crops and removing the nutrients in the crops and everything from the soil. For me, it is all about feeding the soil and also creating habitat for beneficial organisms and a healthy ecosystem in general. Not just for the crop, but for animals and insects." Beth Hartle understands that farming organically has major benefits that reach beyond the farm. She notes that conventional farms pollute the common ground water, but because she farms organically she "is not polluting the land and the air and everything. We are fostering an environment where other things can be living and a diverse and relatively healthy ecosystem. These are all benefits to the greater community."

Most women we interviewed also emphasized other dimensions of the definition of sustainability. Even those farmers who report that their primary motivation is environmental are very conscious that their farm is a business and that the concept of sustainability includes an economic component. Dara, who operates an organic vegetable farm with her husband, elaborates on this more personal aspect of the business,

> Sustainable means it is good for the earth. It is good for the birds in the air, and the water and the plants and the animals that grow here are healthy. It provides a living wage for the farmer and should be able to grow and provide wages for the farmer's kids. Opportunities. There should be lots of opportunities. Sustainable means it can continue. It is

> regenerative. Ongoing if you will. For us it means all of that. It means not going into debt. Not risking losing it because then it is not sustainable. It means being able to say to the lady who comes into the market with an attitude, "I can get this tomato cheaper." It means being able to look her in the eye and say, "And why do you think I do not deserve a decent living?" Yes, to me that is sustainable.

While many of the farmers express a deep commitment to ecological practices, they also recognize that there may be trade-offs. Marilyn runs a small-scale diversified livestock farm but also works off-farm, which sometimes limits the amount of labor and time she can give to the farm. She sees ecological sustainability as a goal of what her farm is moving toward:

> I am interested in the connection of sustainability in terms of practices, ecological farming, and that kind of thing. It is a goal that you aspire to as opposed to something that you are. I look at how the land is used every day, and I see something I do not like about how the land is used and things that I cannot control or just management things that I wish I could do. That is one of the frustrating things for me partly of not having enough time. You know, I am forced sometimes to do things in a way that I would not do if I had more time.

She bought her farm from a conventional farmer, and when she asked him to tell her everything he sprayed on the farm the year before, it took her "three pages to write it down."

> So I know that when I got it and changed it, it has not had any of that on. I know that. I have no interest in using chemicals of that kind. I am interested in looking at maximizing the ecological processes on the farm and as much as you can. . . . You cannot be too doctrinaire about it because things happen, and you are forced one way or another to do things that you would not normally want to do.

Several farmers recognize this tension between environmental quality and economic realities. On reflecting on the sustainability of her farm, Linda Underwood comments,

> I think sustainability is two-fold. One is, is it ecologically environmentally sustainable? Do you do what you can to put more back into the soil than you take out of the soil? So is there an environmental integrity that holds true? Is there an economic reality that is in alignment with your environmental integrity? . . . Can you do good work and also be economically sound? That is a balancing act, and I would image it is a balancing act for most farms. Who you want to be and who you are, you try to make them as close together as you can. I am aware of a very slippery slope that we have with respect to all kinds of things. . . . Where we take shortcuts. Usually the reasons are either economic or environmental . . . I would consider us sustainable. We have been around long enough that we are at least still in business.

Lucinda's idea of sustainability also includes social and personal aspects.

> What I have been thinking about a lot since I started doing this is if it is socially sustainable, but also personally it is sustainable, which is hard I think because I think farmers work way too hard and make way too little money in general, and so I think that is another aspect of it that we need to pay more attention to. How to make it more sustainable from a personal standpoint so people do not get totally burned out.

Lucinda learned that a commitment to environmental practices was not enough if she could not economically and personally sustain her operation.

Women farmers report an interest in sustainable practices for multiple reasons, including respect for the inherent integrity of natural ecosystems and a feeling of responsibility to customers, family members, community members, and future generations. They describe concerns about the current food system, a system that is chemical- and energy-intensive and results in the waste of food and natural products. Their concerns reflect research that suggests about one-third of food produced for human consumption is wasted in the US, and when put in landfills, the wasted food results in methane emissions that contribute to global climate change (Gustavson et al. 2011). For the Perrys, who operate an organic herb and vegetable farm, the sustainability of their agricultural operation is connected to their commitment to reduce food waste and to use renewable energy. Diane Perry subscribes to a "whole food philosophy" that drives their farm.

> Our goal is to eliminate as much landfill waste as possible. . . . So eating very simply and raising foods that are grown in a sustainable fashion using no synthetic, manufactured herbicides or pesticides, or fertilizers. Treating the land as gently as possible and having minimal impact on the land. You know, making an exchange. I mean God gives us so much. We want to be able to return that favor in the way we operate at the farm, and then we try to pass that on to interns, our own family members and our own children and anyone who comes to the farm.

A number of women farmers are committed to reducing energy consumption, especially the use of fossil fuels. As Lori Noble notes, "We are not highly mechanized, we are not expending a lot of fossil fuels." Others, such as Carol and Sasha focus on sustainable methods of building. They built a straw bale building to use as a classroom, a distribution center for their CSA, and a processing area. They also plan to install solar panels and put in a composting toilet. Carol claims, "We want to model all of those things as a holistic kind of circle of living."

Nan, an organic vegetable farmer, sums up her views on sustainability:

> Well, sustainability partly. It means that we are farming in such a way that we can continue to farm. That the fields do not get depleted, and we are not harming the environment. . . . Sustainability to me means that we are making enough money that we feel that we have an acceptable standard of living. It does not mean just getting from day to day. In fact the farm is supporting us and can support something in the future.

Debbie Thompson approaches farming by thinking of future generations in all that she does. As she explains, "Oh I think for our future generations. We are just borrowing this land for the future. We are here and we are borrowing it like our ancestors before us. We have got to make sure that it is viable for the generations to come. Can't people see beyond that? I think it is very important to save farms. I think the land is just so precious."

Together these women's definitions reflect a complex and multidimensional view of sustainability and how those definitions are manifested on their farms, in their products, within their families, and in the environment. Their descriptions recognize trade-offs among the multiple goals they have for the farm, but a striving to keep moving in a direction—toward greater sustainability—they feel is right for them, their families, and their communities. As Linda termed it, this "balancing act" between their aspirations and the realities of their lives becomes expressed through their farms and through the food system.

In the following two sections, we delve deeper into the meaning of sustainable agriculture to women farmers in our research. We do so by describing how and why women farmers began farming sustainably, as it offers an important window into how they enacted their values through their farms. We have identified two main pathways by which women farmers have come to sustainable production on their farms: those who transitioned from conventional to organic production, and those who entered farming with the intention of using environmentally friendly practices. We provide examples of both pathways into agriculture and discuss the implications of their choices.

Making a Change: Conversion from Conventional to Sustainable

Farmers transition from conventional to organic or other "alternative" production systems for various reasons, including personal health, economic drivers, and concern for the environment. Here we provide three cases of women who converted their operations from conventional commodity production to alternative production strategies: Natalie Ingram, a feed grain farmer; Elaine, a hog farmer; and Karen, a dairy farmer.

GOING AGAINST THE GRAIN

Natalie Ingram and her husband, Larry, were growing grains (primarily corn and soybeans) conventionally with his brothers. Natalie and Larry were typical of most conventional grain producers. Corn is king in the US, with corn grown for grain accounting for almost one quarter of the harvested crop acres on over 400,000 farms in the US (USDA 2013b). Soybeans rank second just after corn, grown on approximately 280,000 farms, and the acreage devoted to growing soybeans is very similar to that of corn grown for grain (84 million acres in 2011). The majority of corn and soybeans in the US are produced using chemically based herbicides, insecticides, and fungicides (USDA 2013c).[4]

Both Natalie and her husband Larry fashioned themselves as experts in chemical agricultural production. She worked as a scientist at an agricultural experiment station where she helped farmers develop spray programs for their crops. She describes Larry as "the local expert if anybody needed a cocktail to control a particular weed. Everybody in the neighborhood came to him for advice on pesticides because he knew more than the extension people did." She jokes, "One of the things that we spent a lot of time talking about when we were courting was this chemical and that chemical. You know we were into this. This is part of who we were."

When the brothers decided to divide the farm, Larry and Natalie felt that five hundred acres of conventional corn and soybeans was not enough to support their family. Natalie says, "So almost immediately we started looking at various niches that we could possibly go into." They considered various types of value-added processing, such as raising heifers. After considering different scenarios, they realized that many of their ideas would require large investments, and others were quite risky. Natalie recounts the time they first considered organic production but considered it to be too risky: "About that time, we were looking through the local farm paper and saw an ad that said, 'Wanted: organic wheat.' We looked at each other and said, 'We know that would never work.'"

According to her, the event that spurred their transformation from chemically intensive farmers to organic farmers was when Larry became sick from spraying chemicals.

> [Larry] . . . wore one of these white zoot suit things when he was spraying, and he came in and he would stink. It was awful. . . . He had been spraying weed spray and he came in feeling really bad. I mean really lousy. Sick to his stomach and just not feeling well. He went to bed early and woke up the next morning, and his whole right side was paralyzed. It was scary, and we were right in the middle of hay season, and we had to get hay out of the field, and fortunately we had good neighbors, and so they and I unloaded all of that hay, and [Larry] just stood there with his arm hanging, because he could not move his arm. We went to the doctor, and here in the middle of June in farm country, and never once did that doctor ask, "Have you been spraying with any neurotoxins?" He gave him muscle relaxers and pain pills and stuff like that. . . . Eventually he got back his

> strength, but it never came back all of the way. About five years after that, I was rereading *Silent Spring*, and Rachel Carson described 2,4-D poisoning and to the tee. He was poisoned. He was poisoned by 2,4-D, and you know, nobody bothered to diagnose it or do anything about it.

They transitioned from conventional to organic grain production and now manage over 1,400 acres of organic grain. They have become leaders and promoters of organic agriculture through encouraging their neighbors to move to organic production and purchasing and operating an organic feed mill in their town. The feed mill provides a market that supports organic grain producers by providing a place to sell their grain and supports organic livestock producers by providing them with a source of organic feed.

FROM PIGS IN PENS TO PIGS ON GRASS

Elaine's husband worked for a number of years on confinement hog farms. Over the last fifty years, hog operations have become larger and more concentrated, and farmers typically work under contract with pork processors and specialize in a single phase of production. Hogs in these specialized operations are grown in confinement at high stocking densities as part of a systematic effort to produce the highest output at the lowest cost (McBride and Key 2013). Elaine describes the difficulty of working in a confinement hog facility: "The animals are housed totally under roof and above slatted floors. The odor level is real bad. The dust level is horrible. You have to wear dust masks. It is very hard to get used to using them. Some people do not use them. Therefore there is the dust particle issue, and it is just a rough environment. "

Elaine and her husband started pasturing their pigs, she says, "mostly because it was a lower cost and because we just preferred to be outdoors, but a lot of it, more so now, has to do with more of the sustainability issues." When they first started out over a decade ago, she explains, "[We were] still conventional farmers doing sustainable ways, but not really knowing what sustainable really was. . . . It was purely personal preference. It was purely money. I feel like we have come full circle to understand really what we were doing." Although she does not personally take credit for the conversion from conventional confinement to pasture-based production, she notes, "I am a little bit more open-minded than my husband. He is a little bit more of one mindset. For it to even have overtaken him, it is really amazing." She adds,

"We have been at it for eleven and a half years now. . . . We knew it was healthier on the pigs from the get go. We did not put antibiotics in the feed for cost reasons. We were doing a lot of sustainable things, but did not know that we were doing it."

Although their outdoor production system was working, they found that they were caught in a cost-price squeeze due to high feed costs and low and fluctuating hog prices. They made the decision to stop selling their hogs on the conventional hog market and to market directly to customers. As a result, they have succeeded in stabilizing their profits while at the same time downsizing their operation. As Elaine happily describes their current operation, "We used to be three times bigger than we are now. We are down to sixty sows, so our goal is to direct market everything that we have." By downsizing, adding value to their pork by producing on pasture, and direct-marketing, Elaine and her husband are capitalizing on marketing meat produced in ways that benefit the environment.

MOO-VING FROM CONFINEMENT TO GRASS-BASED DAIRY PRODUCTION

Karen and her family raised dairy cows in a conventional confinement system. In confinement dairies, cows are housed in barns, milked in parlors, and fed mechanically harvested feed all year. Production per cow is maximized to offset the high investment per cow and low profit margins. The milk processing industry favors large confinement systems because they produce larger volumes of milk on a consistent basis. About 15% of dairy farms in 2007 used recombinant bovine somatotropin (rBST), also known as recombinant bovine growth hormone (rBGH) or artificial growth hormone (USDA APHIS 2007).[5] Use of rBST is associated with an increase in milk output ranging from 11% to 16%, but also a 25% increase in the risk of clinical mastitis (infection of the udder), a 40% reduction in fertility, and a 55% increased risk of developing clinical signs of lameness (Khanal, Gillespie, and MacDonald 2010). These risks can lead to increased expenses and smaller profit margins, and they possibly contribute to an overall increase in the scale of dairy production in the US.

Karen and her family, like many dairy farmers, found that they were no longer profitable because of the high costs of feed, especially when a drought led to drastically increased grain prices. They decided to transition to a grass-based dairy. In grass-based dairy systems, dairy cows graze on managed grass and legume pastures instead of being confined indoors and fed hay, grain, or cut forage. Cows in confinement operations generally produce higher milk yields than those that graze, but production costs and labor are generally lower in grazing systems. Because fewer grain crops are grown to feed grass-fed cows, graziers generally apply fewer chemical pesticides than those raising grain crops to feed their cows.

Karen and her family now milk 275 dairy cows using a grass-based system. Karen is pleased about the transition for a number of reasons: "So we have been grazing now since 1994, and it has made the biggest difference on the farm. We enjoy it and it is beautiful. It just seemed like after we did that, our farm was just so much more profitable and fun." She has noticed that the health of their animals is improved: "Our cows just last longer. You know they are outside and they are healthier." They also have refrained from using recombinant bovine growth hormone on their cows because of the health risks to the cows, health risks to people, and costs.

> The life span of a lactating cow now on those big conventional farms is two to three lactations and they are gone. That is one reason why cow prices stay so high is because there is always a shortage on these big farms for their replacements. I mean we just do not feel that we want to do that. We do not want to produce milk that has that in it. The other reason is, it is so costly. The cows do not do as well. The longevity of their life is shortened by using something like rBST. Farmers cannot get cows bred when they are using that.

Another major benefit from their switch from confinement production to grazing is that their workload has decreased because they do not have to harvest as much hay, handle manure, take the risk of growing crops, or invest in equipment. Explains Karen, "Another thing is you are not processing all of that hay and bringing it into the farm and storing it. You know you are letting the cows go out there and chew it off and eat it. . . . She is doing the work. . . . Then the other thing is, the manure handling. . . . You do not have to do all of that manure handling in the barn and storing it and spreading it, and that has been a source of savings." Karen has noted significant improvements in their profitability by switching to a grass-based system:

> I think another thing is we were putting all of that money into crops. We would take a loss sometimes with the drought. . . . Another savings was equipment and gas. We are just not doing equipment. Equipment is really a lot of farmer's problem. . . . They are getting so much expensive equipment that . . . for the commodities and the products that they sell, they just do not get [the investment] back. You buy a seventy-five-thousand-dollar machine and it depreciates and rusts, and you know, it is a bag of bolts someday.

Karen sums up the benefits of their conversion to grass-based dairy production.

> I would say we are sustainable because we were not sustainable before. We were working and working and working, and you know our land was eroding because we were plowing it up, and now our land is so much more valuable because it is organic. It has been cared for in a much better way. We are sustainable in that we are surviving when a lot of farms are not surviving. I mean all of our neighbors here, if I talked to those women it is just doom and despair. They . . . cannot pay bills. They are not getting enough money for their products.

By emphasizing sustainability—in all of its dimensions—Karen and her family have been able to buck the statewide trend that shows a significant exit of dairy farms in the past decade (Center for Dairy Excellence 2012).

Beginning Farming with Sustainability in Mind

In interviews, a number of women farmers in Pennsylvania reported that they began farming with a keen interest in environmental issues and a larger vision of sustainability. They described concerns about the negative environmental impact of agriculture on the environment, including loss of natural habitat, pesticide use, and soil erosion and nutrient loss from land. Nina is one example of a woman who started farming because of her interest in environmental quality.

> My reasons for . . . switching to working in agriculture—it came from an interest in environmental conservation. I was working in conservation for

> the National Park Service and feeling that National Parks are great and I love them, but they are not always accessible to everyone. So by working in sustainable agriculture I would be working in an area that could affect everyone. Because everyone has to eat, spreading the message of environmental conservation in a way people can do it in their daily lives. So I definitely came from the environmental perspective. Through doing an internship I just learned a whole lot more reasons to do sustainable agriculture and to promote sustainable agriculture.

Dara Young began farming because of her recognition of the health consequences of chemically intensive agriculture, particularly a concern about the link between pesticides and cancer. Farmers and their families are at the highest risk from exposure to agricultural chemicals, which include pesticides and synthetic fertilizers. Many agricultural chemicals are known or suspected of having either carcinogenic or endocrine-disrupting properties. Exposure to these chemicals has been linked to multiple disorders and cancers. Because agricultural chemicals often are applied as mixtures, it has been difficult to clearly distinguish cancer risks associated with individual agents (President's Cancer Panel 2010).

As a nurse, Dara and her nursing friends have seen many people die from farm and garden chemicals. As she reports, "Friends that I knew who worked in ICU see people die of cancer. A guy sprays his fruit trees and got a back spray on him, and he died of liver failure within three days." These incidents propelled her into farming. "I just decided I wanted to grow my own food. Like my personal family cancer rate is really high and so is Kevin's. Really high. We have had every cancer and every risk factor. So we decided that we were going to start to grow our own food."

Many of the farmers expressed similar concerns about the problems of chemically intensive agriculture and wanted to farm in a different way. Debbie Thompson is committed to farming without chemicals. "I do not want any chemicals. We have a creek going around the place. I was not going to be responsible for the runoff that went into that creek. My farm started out with fifty-two acres. Then the neighbor's farm, my goodness they used to lease it out to somebody, and the spray would float down the creek. I lived down at the bottom in a valley, and I could not breathe at night." She looked up what chemicals they were spraying and concluded, "The chemicals they were spraying were killing us." Even though she was in her sixties, she decided to buy the neighboring farm to stop the spraying of chemicals and the pollution of her creek. She farms organically and claims that "after World War II, the chemical companies came into being, and that was the worst thing that ever happened to us." The stories from these women suggest a growing desire to enact values of sustainability by intentionally moving into farming so as to enhance the quality of both human and environmental health.

Conclusion

The innovations that women farmers have developed—innovations that help them achieve success on their farms in a way that satisfies their personal values related to community, sustainability, and providing for their families—can also be seen as providing the foundation for an alternative food system. For example, Linda's farm, and the multiple income streams she has built, is a year-round enterprise that provides an opportunity for customers to connect to the food system and to their community. Her farm is "trying to provide information, education, inspiration for the home gardener. We have tools for people to grow organically. We get lots of different seed sources for the cook and gardener—interesting varieties, heirloom tomatoes, lots of herbs and that sort of thing. . . . We also sell a lot of local-made crafts that provides a market place for many local artisans." Linda's farm has become an important place for community members to connect to food and the earth:

> I think of it as the agricultural conscience of the community. As the area continues to develop and the homes that are coming in this area are over a half a million dollars if not $600,000 and $700,000 homes. As the agricultural landscape is disappearing and now farming becomes less and less of a vocation, I think that part of what we represent both physically and maybe emotionally, psychologically, and spiritually. . . . It is like maybe we provide some connection still to . . . some sort of a simpler life. It is not a simpler life, but maybe it is a less complicated one. So maybe two steps closer to the land and the earth. We continue to provide the information and inspiration about that. About food. About seasonality. About plants and people's gardens. Things that are more central to the core, to our hearts, to our bodies. I think that is probably who we are in this community.

Linda's business has pushed to create and capitalized on the growing interest in local foods. By working with others in the community, "we have been able to form our own food system in conjunction with farmer's markets and restaurants."

Bringing people to the farm not only grows interest and demand for local, sustainable products; it also may lead to greater awareness of health issues associated with our conventional food system. For example, Dara notes that she has witnessed a substantial change in their customers' concerns over the years, even among a rural and less wealthy community:

> I think people are more aware. I think the big ag things are scaring people—the strawberries a few years ago and the spinach and bovine growth hormone. I think people are more aware. I have young couples who are not necessarily wealthy who come in the driveway, and they say they do not want to feed [their] kids that stuff. What is that? I want to know where it came from. Also, people like to have their kids in touch with something outside. So we have some paths mowed down for the woods, and they can go down there and they can go and see the chickens and step in and pick up an egg, and so that means a lot to people with small kids. We have every age range coming. I believe people are more aware. We have been doing this a long time. I think that the awareness is different now.

For Linda and for Dara and Kevin, the farms they have built—through their own innovations and adaptations to the existing agricultural systems—have provided the chance for people in their community to connect to their farm, their landscape, their food, and others in their community. This may signal the foundations of a movement toward a new farm and food system, a system that builds on the innovations of women farmers as they have sought to develop their farms' businesses in concert with their own values, which reflect the multiple dimensions of sustainability and emphasize creating successful businesses that sustain their families, building community with farmers and consumers and seeking greater balance with the ecological systems that provide for human, plant, and animal health.

This new farm and food system is consistent with the local agriculture movement. Although the goals of the local food movement vary, those who advocate for local food see it as a way to develop food security, economies, and communities in a specific place rather than through industrial operations and corporations elsewhere (Feenstra 2002). Although the definition of *local* varies, there are many indicators that there is a growing interest among consumers in eating food grown locally (Martinez et al. 2010). For example, the number of farmers' markets grew threefold between 1994 and 2010, and local foods are becoming highlights of many restaurant menus. Direct sales through CSAs and farmers' markets grew by over 100% in the last decade and a half compared to 48% of all agricultural sales (Martinez et al. 2010). The local food movement has contributed to the growth of what Lyson and Guptill (2004) refer to as "civic agriculture"—farming that builds the local social and economic community as well as producing food. This movement emphasizes a shared commitment to building a more equitable agricultural system, one that allows growers to focus on land stewardship and still maintain productive and profitable small farms. Women farmers are at the forefront of this movement, taking advantage of the growing interest among consumers in eating local food and building local community relationships that are intended to further grow the movement.

In addition to building a new food system, the use of strategies such as value-added production, collaborative marketing, and on-farm education also signals a redefinition of what it means to be a farmer and what farmers do to incorporate revenue generation opportunities that draw from traditional women's work. Women's traditional responsibility for reproduction, food preparation, caring, entertaining, education, and community building have been transformed into valuable labor and a source of profit for farms. These changes also indicate a blurring of the boundaries around traditional gendered divisions of labor on farms. Women, as farmers, are taking on farm production roles but drawing from their personal—and professional—identities as women.

Chapter 5

Constructing a New Table: Women Farmers Negotiate Agricultural Institutions and Organizations, Creating New Agricultural Networks

Educational and social organizations are critical in empowering farmers to gain access to resources, in building rural social capital, in helping farmers to gain skills, information, and markets, and in increasing farmers' voices in public sector decision-making (Penunia 2011). Many women involved in agriculture find that the established avenues, organizations, and networks that male farmers use to access educational and technical resources are less accessible or even irrelevant to women farmers (Hassanein 1997). While there are many agricultural organizations and institutions that could help women farmers gain access to the information and resources that they need, many organizations have not fully accepted women as farmers and have not sought to advance gender equity in agriculture. It is this need—this gap in information critical to their business success and social legitimacy as farmers—that has led to the development of organizations and networks that specifically support women farmers.

In this chapter we first describe the needs of women farmers that they expressed during research we have conducted, as well as those reported in the work of others around the country. We then juxtapose these needs against the approaches of existing organizations, particularly Cooperative Extension, to highlight the lack of services and support for women as farmers and as women. In response to the lack of adequate organizational support, women farmers have created agricultural networks (including the Pennsylvania Women's Agricultural Network, PA-WAgN) that are intentionally focused on providing access to the resources they need to successfully overcome the challenges all farmers face, as well as the challenges women specifically face. These networks offer access to knowledge- and skill-building opportunities, new ideas from other women farmers, and social support that can alleviate isolation, legitimate women's identities as farmers, and increase their capacity as farmers, resulting in the enhancement of their farm businesses.

Identifying the Gaps: Women Farmers' Educational and Networking Needs

In the early stages of the development of the Pennsylvania Women's Agricultural Network (PA-WAgN), a group of faculty and graduate students acquired funding to conduct research to identify women farmers' educational needs and to develop programs to meet those needs. The emphasis on systematic qualitative and quantitative research as a guide to organizational structure and programming is a strength of PA-WAgN. In the initial stages of this research, women farmers participated in focus groups across the state to discuss their educational needs and the factors they felt enhanced their success. In this initial research, women farmers revealed they struggle to be recognized as farmers; feel isolated; lack skills, land, and capital necessary to operate their farms; approach problems differently from men farmers; and desperately want practical information and education about farming (Trauger et al. 2008; Barbercheck et al. 2009). These research findings guided decisions about how to structure PA-WAgN and how to shape educational programs. PA-WAgN continues to use surveys, evaluations, needs assessments, in-depth interviews, focus groups, and ideas from regional representatives to learn what issues concern women farmers, how to design innovative agricultural programs and market them, and how to define the direction of PA-WAgN programs. In this section we highlight the results of this research to lay the foundation for understanding what it is that women farmers need (and what they are not receiving) from the educational and technical support system for farmers.

Problems within Agriculture's Culture

A key finding of PA-WAgN's research is that women farmers experience problems within agriculture's culture that impedes their success as entrepreneurs by limiting access to the information they need. In initial exploratory research, women farmers identified a list of potential challenges to being a woman farmer. We later used this list to develop a formal needs assessment survey conducted from 2006 to 2007. In this survey, almost two-thirds of women farmers reported that they were not taken as seriously as men farmers, and about half reported not being welcome in agricultural groups. Additionally, about half reported facing a lack of family support for their role in managing their farm. Their ability to farm successfully, they said, was also tied to their isolation from other farmers (in particular women farmers), their lack of a farm background, and for some, and their lack of childcare (Barbercheck et al. 2009). These cultural and structural problems have implications for the ability of women farmers to obtain agricultural information, obtain respect, achieve leadership within the agricultural community, and conduct agricultural business profitably.

Educational Needs and Learning Environment

In PA-WAgN surveys, focus groups, and interviews, women farmers expressed interest in a broad array of educational topics, which reflect women's multidimensional motivations in running farms that often have diversified businesses, products, and systems. The 2006–2007 needs assessment reflects these topics. Fully two-thirds of the women farmers wanted to attend educational events to increase crop or livestock productivity, increase soil fertility, and manage pests, but also to manage finances and to build infrastructure on the farm. About half wanted to learn how to work with the local government to represent farming interests and to maintain environmental health. A majority also expressed interest in mechanical skills, such as how to operate and maintain farm equipment, and business skills, such as how to market their products, whether or not they grew up on a farm. Women farmers reported the least interest in attending training related to family and parenting issues (Barbercheck et al. 2009).

Women farmers overwhelmingly express interest in having educational programs designed in very specific ways to enhance their ability to learn in the limited time they have available. They want participatory, hands-on, and peer-led programs, drawing on the knowledge and experience of a network of women (Barbercheck et al. 2009; Byler et al. 2013). Many women farmers in our first focus groups reacted negatively to their experience with existing, nonparticipatory formats in agricultural education. These formats were expert-led presentations using the banking model of education, a metaphor

Terra Brownback of Spiral Path Farm demonstrating greenhouse tomato trellising. Photograph by PA-WAgN staff.

used to describe education as educators depositing information into students who are empty vessels (Freire 1970). Instead, many women farmers expressed their desire to learn in a participatory environment. They criticized other types of educational settings for lacking interaction and for not recognizing the social relationships that are essential for learning to take place. For instance, one said, "PowerPoint . . . just really wrecked a lot of things. Many, many [instructors] are not really versed in how to speak using PowerPoint and keep your audience. So you attend many, many talks . . . where they are canned. You know [instructors] go out and do the same talk over and over. You do not get anything out of it." Another participant adds, "PowerPoint should be shot as far as I am concerned." These canned presentations are not tailored to the needs of the audience, assume a lack of expertise by the audience, assume an expert role by the instructor, and are not often used to promote participation and peer learning opportunities.

Most women reported that they learn best from other women farmers. They emphasize what they consider the different learning and teaching styles of women and men. Many women farmers look to both men and women farmers for information and encouragement, but like Irene, they made a distinction about differences in what they received and sought from men and women farmers.

> I always feel like I can get good technical information from male farmers. I feel when I speak with other women who are doing the same thing I am doing, that it is more nurturing. The information is shared in a different way. Even if the information is not shared, it is the camaraderie. Again, I get something different out of speaking with the women than I do with a male. It is just more than technical information.

Our research and interaction with farmers continually reinforces the idea that hands-on peer-to-peer education works best. Carol explains that given her off-farm job and demands of the farm, she finds it very difficult to take the time to attend educational events, but she is willing to take the time to attend field days led by farmers.

> You may even just get like a little spark of an idea from what somebody else is doing for you to try something different. It makes the biggest difference in what you do. So I think field days are an asset for me because I am a very hands-on visual person. That for me is the absolute best way.

Irene, like many other farmers, emphasized that she learned best through hands-on experience. She explained that she goes to many conferences and workshops and uses this information to go back to her farm and try what works for her. But she emphasized that something is missing for her in this type of education.

> That is one of my issues with conferences and workshops. I need hands-on stuff for me to truly learn something. . . . It is just my learning style; then I really learn it better when I do hands-on stuff. It is like if I drive somewhere. . . . If somebody drives me some place, I would probably have more of a challenge to try and get back there, but if I drive I could get back there with no problem.

In summary, women farmers who have participated in PA-WAgN research prefer interactive, participatory, and flexible programming that meets their technical and educational needs.

One reason that women prefer these learning formats is that they do not want to feel intimidated. Compared with men farmers, women have different farming and educational backgrounds, knowledge bases, and physical abilities, which often means they ask different, more basic questions. Asking such questions can be difficult to do in a setting with men who have more experience and may not regard women as legitimate farmers. The physical differences between men and women are particularly acute, and have been mentioned frequently by women farmers in our research. Women emphasize their different bodily strength and experience with farm equipment. For example, Marilyn explained that she values "learning from women and how women do things, versus learning from a man who does the same thing in a manly way." She gives the example of working with a female intern who could show her how to deal with horses in a way that Bob, the man she farmed with, could not: "[He] could not show me because he is a big man, and he just did it his way." Another woman farmer agreed and expressed a desire for having a network of women who can give women farmers advice on how they perform tasks, not only because men tend to be different in bodily strength and have previous experience with equipment, but for other reasons: many women farmers lack resources and basic machinery skills. She said that her male farmer neighbors do things a certain way. "I can't do it the way they do it. I am not as strong as they are, and I do not have the same resources or the same mechanical knowledge, you know. I don't really even know how to change the oil in the rototiller."

In summary, PA-WAgN research on women farmers in Pennsylvania and the northeastern US has consistently found that women farmers prefer educational formats that allow peer-to-peer exchange of information, hands-on and interactive activities, the ability to learn from those who have experience with a particular practice (especially from women), discussion of multiple criteria for judgment of success, and informal discussion (Trauger et al. 2008; Barbercheck et al. 2009; Byler et al. 2013). In these educational settings, farmers access and assess locally specific knowledge and adapt it for their own purposes (Hassanein 1997; Carolan 2006; Jordan et al. 2003). These qualities, however, do not typify most educational or other events sponsored by traditional agricultural organizations. We turn next to describing the approaches of these traditional organizations, to highlight the ways in which women's needs were not being met by them. In fact, it is their experiences with these traditional approaches by agricultural organizations that have pushed women farmers to start their own organizations.

Finding a Place at the Table: Women and Traditional Agricultural Organizations

In this section we briefly review women's involvement in and experiences with organizations that serve and advocate for farmers and agriculture in the US and the (limited) role that these organizations have played in serving the needs of women farmers. We focus on two main types of organizations. First, we pay particular attention to Cooperative Extension, which is the established public institution of education and support for farmers in the US. We analyze the ways Cooperative Extension has incorporated women as an audience over the years and the experiences of women farmers with Cooperative Extension services. Next, we focus on general farm organizations, such as the American Farm Bureau, National Farmers Union, and the Grange; commodity organizations and their affiliated women's auxiliary groups, such as the Porkettes and CattleWomen; women-only organizations such as Women in Farm Economics (WIFE) and American Agri-Women; and sustainable agricultural organizations, such as the Pennsylvania Association for Sustainable Agriculture (PASA) and the Practical Farmers of Iowa. We argue that Cooperative Extension and some of these other agricultural organizations have not been meeting the needs of women farmers because they often define women only through their relationship with the male farmer, and not as women farmers in their own right.

The Cooperative Extension System

A HISTORY OF COOPERATIVE EXTENSION PROGRAMS FOR MEN AND WOMEN

National public interest in supporting agriculture has resulted in an integrated system of research and educational support institutions that include land-grant colleges, agricultural experiment stations, and Cooperative Extension. Land-grant colleges were established by the Morrill Act in 1862 in every state to provide education in agriculture and the mechanical arts. Each state received grants of 30,000 acres of federal land for each member in its congressional delegation, and when sold by the states, the proceeds funded public agricultural colleges. Each state established a college, and every state continues to have a land-grant college to the present day. In addition, some southern states established through the Morrill Act (1890) separate land-grant colleges for blacks (Comer et al. 2006). Agricultural experiment stations, usually affiliated with the land-grant colleges, were established in every state by the Hatch Act of 1887, and they support research that focuses on agricultural productivity and technology relevant to their individual states.

We focus mainly on Cooperative Extension, as it is the main system for providing educational and technical support to farmers. The Smith-Lever Act of 1914 formally established the Cooperative Extension System in the US to form a partnership between the US Department of Agriculture, land-grant colleges, and state agricultural experiment stations to extend research findings from these institutions to citizens. The mission of the Smith-Lever Act was "to aid in diffusing among the people of the United States useful and practical information on subjects relating to agriculture and home economics, and to encourage the application of the same" (Conglose 2000). When Congress established Cooperative Extension, 50% of the population lived in rural areas, and 30% of employment was in agriculture (USDA 2015b). Cooperative Extension was designed to meet the nonformal educational needs of rural and farm men and women, particularly those who could not come to the land-grant colleges for formal education. The intention was that federal support of farmers through Cooperative Extension would both increase agricultural production and keep people on farms to sustain a food supply. Through partnerships between the federal government, states, and local counties, Cooperative Extension established offices in every county of the United States to provide scientifically based education programs on agriculture and home economics.

The agricultural research-Extension link typically involves the development of new technologies and practices by researchers at land-grant universities, followed by the dissemination of this information by Cooperative Extension educators, with the intended goal of adoption of these technologies and practices by farmers. The basic assumptions in this expert, top-down model of education are that farmers will adopt what the researchers develop and that the new technologies are appropriate for all farms (Colasanti et al. 2012). Cooperative Extension takes credit for spreading research results and advising farmers, thus contributing to an agricultural revolution over the last century that enabled fewer farmers to produce more food.

Cooperative Extension, as a publicly supported educational institution, perceives itself as an essential contributor to national goals of food security and rural economic development. How then, has Cooperative Extension served rural women and women farmers? From the outset, Cooperative Extension had different sets of programs for men and women and separate educators to teach them. Historically, male Cooperative Extension educators addressed agricultural information needs of farm men, while female Cooperative Extension home economists taught nutrition, food preservation, gardening, poultry production, farm safety, and sewing to farm women. The American Home Economics Association, formed in 1908, promoted home economics education for women and girls, lobbying federal and state governments. The Smith-Lever Act of 1914 specifically focused on home economics education for adult women, resulting in a rapid expansion of educational programs for them (Heggestad 2015). Similar to the focus on science and economics for improving farming, home economists sought to emphasize how principles of science and economics could increase the efficiency of women's work. The adoption of new technologies in the home encouraged more urban middle-class lifestyles for farm women and, perhaps, the subsequent removal of some farm women from agricultural production activities (Trauger et al. 2010; Neth 1995; Jellison 1993).

Cooperative Extension had a broader educational agenda than just the dissemination of new technologies, however. It also helped farmers, typically men who specialized in particular commodities, build networks and organizations. It helped farmers to form local commodity organizations to share information about yields, pests, and diseases affecting production of their particular crop or livestock. Cooperative Extension offered leadership training to local farmers and commodity groups, training farmers to be the leaders and spokesmen of agricultural interests (Black 2007). Commodity organizations under the direction of these spokesmen and agribusiness interests influence the direction of agricultural research at the land-grant universities and the direction of educational programs at the local and state levels. They also provide the political support necessary for continued federal and state funding of agricultural research (Middendorf and Busch 1997).[1]

Cooperative Extension has evolved to provide a broader range of programs to both rural and urban audiences with the result that today there are six major Cooperative Extension program areas: agriculture, family and consumer sciences, 4-H youth development, natural resources, community and economic development, and leadership development. While these areas are no longer purposely divided by gender, the major programs in agriculture have tended to serve men clientele, emphasizing marketing, management, and productivity, while the area of family and consumer sciences has tended to serve women clientele, teaching nutrition, food preparation, child care, family communication, and financial management. The one exception is that food safety programs have drawn both women and men because of state regulations for restaurant kitchen workers.[2]

COOPERATIVE EXTENSION EDUCATORS' INTERACTION WITH WOMEN FARMERS

Women farmers with whom we have interacted have described mixed experiences with Cooperative Extension. To understand how current Cooperative Extension educators perceive women farmers and if and how Extension educators work with women farmers, we employed two strategies: in-depth personal interviews with agricultural Extension specialists in Pennsylvania followed by an online survey of Extension educators (Trauger et al. 2010; Brasier et al. 2009). Overall, the participants revealed considerable differences in the extent to which Extension educators interacted with women farmers: some reported little or no interaction and others frequent interaction. Some Extension educators disclosed in the interviews that they did not consider many women living on farms to be farmers, but rather consider them to be helpers, off-farm workers, or bookkeepers. Other Extension educators recognized that women farmers were not taken seriously or not welcomed by other farmers in the agricultural community. An Extension educator explained his experience in working with farmers, "I was amazed at the things [men] farmers said. They really objected to women getting involved with production agriculture, as owners or managers. I really got the sense that they felt women didn't belong."

Women farmers have articulated specific educational needs (Barbercheck et al. 2009), but women farmers have also reported difficulties getting these needs met through Cooperative Extension. Consequently, we wanted to understand how Extension educators perceived women farmers' needs and if those needs differed from men farmers. In this research, just over half (58.7%) of the educators recognized that the educational needs of women farmers are somewhat or very different than men farmers; however, one quarter (26.6%) reported no difference, and about 15% stated that they had never considered that there might be a difference (Brasier et al. 2009). This finding suggests that a sizeable segment (41.6%) of educators surveyed might not be sensitive to the possibilities of differing needs of women farmers.

To further explore how educators perceive the needs of women farmers, we conducted interviews with Extension educators and statewide specialists in relation to women farmers' needs, including preferred pedagogical approaches and educational contexts. Those educators who believed that women farmers had different educational needs also believed that women farmers had different learning styles than men. Several educators mentioned that women tended to be more vocal and interactive compared to men, who they described as wanting "only the facts"; for instance, "Certainly in a setting with all women, you get more interaction [than in a setting with all men] . . . sometimes that becomes hard to manage because it often gets off topic, and you only have so much time to get through a subject." He adds, however, that women often seem to be more interested and engaged than the men. Open comments in the survey confirm from other educators why women want a different environment; for instance, "The challenge in providing education for females in a mixed audience is that some people who want 'just the facts' may make those who want to explore, ponder, and listen to others' experiences [i.e., the women], uncomfortable. The learning environment is the big factor that needs to be different for many women." Some educators confirmed our findings from interviews, surveys, and focus groups of women farmers that women farmers do not find the predominant mode of Extension information delivery suitable to meeting their needs.

One educator who recognized that some women feel uncomfortable at traditional Extension meetings designed programs particularly targeting women farmers. Although he says that his programs attended by both men and women have been successful, he believes that some women are not comfortable or empowered to participate in mixed settings. He expresses concern that Extension educators lack formal training in how to be adult educators per se, but are merely expected to present canned educational programs. He thinks it would be helpful for Extension educators to understand how adult learners vary from younger learners, and how men learners can vary from women learners.

Despite the awareness of this one educator, several other educators portray a substantial lack of sensitivity to the possibility of differences in what men and women farmers may want in an educational program. These educators emphasized that the content of their programs was not gendered and that they strived to be gender neutral in their classrooms. As one educator stated, "I treat everyone the same. I believe in equality and all that stuff, so much so, that I don't even think about it. . . . I have never been exclusionary, never felt like I need to make blacks or women feel welcome, because they are." Another explained, "I don't feel I have to market programs differently. I feel that the quality program I am putting on will be open to either gender if that's what their needs and interests are." These educators clearly lack sensitivity to historical patterns of discrimination and fail to realize that treating everyone the same does not necessarily result in equal opportunity.

Our experience is that Cooperative Extension in Pennsylvania has made some effort to recognize and serve women farmers, but that it has, overall, been uneven in its response to this growing demographic. In 2005, Penn State Extension supported a half-time staff person for one year to facilitate educational programs for women farmers. County-based educators cotaught a few programs in horticulture, and others advertised programs organized by women farmers if the educator's regional director accepted women farmers as a valid target audience. Cooperative Extension continues to include the number of women farmers at educational events in its civil rights reports. Over the long term however, Cooperative Extension in Pennsylvania has provided no institutional or systematic mechanism for integrating the documented needs of women farmers into its plan of work. As a result, women farmers are not integrated into the system of support that farmers have traditionally received, including design of educational programs, identification of research needs, organization of demonstration plots on local farms, and development of the leadership skills of women farmers so that they may be an effective voice in agriculture.

Here we have discussed the particular case of Cooperative Extension in Pennsylvania, but we must note that Cooperative Extension can operate differently in other states and at the national level. One national Cooperative Extension program that has been developed for women on farms is Annie's Project. The mission of Annie's Project is to "empower farm women to be better business partners through networks and by managing and organizing critical information."[3] Annie's Project was originally developed by Ruth Hambleton, an Extension educator at the University of Illinois, in 2003. By 2013, Annie's Project was offered in Iowa, Missouri, Wisconsin, Nebraska, and Indiana (Hambleton 2013). Since that time, Annie's project has been offered in thirty-four states, and the USDA's Risk Management Agency offers funding for the program in many states.[4] Annie's project recognizes that farm women like to learn in groups and to support each other. The course is usually organized as six educational sessions on topics including financial planning, management, legal issues, marketing, production tools, and other topics chosen by participants. Extension educators receive training on Annie's Project before they offer the course in their states.

In 2008, PA-WAgN worked with several Penn State Extension educators to adapt Annie's Project to reflect topics identified by our research as of particular interest to women in sustainable agriculture. Because the original Annie's Project was developed and conducted in the Midwest, topics such as commodity futures and contracts were of less interest to women farmers in the Northeast, particularly those growing sustainably and for direct markets. Annie's Project for the Northeast instead emphasized direct marketing, value-added processing, and diversified farm products and entailed a more participatory model than the original.

Another federal effort began in 2006 when the US Department of Agriculture sponsored a biennial National Extension Women in Agriculture Conference. This conference brings together Extension educators and other agricultural professionals to share strategies for helping women producers manage financial, production, marketing, and legal risks associated with their agribusinesses. An additional example of the USDA's support for educational programs for women farmers is the Beginning Farmer and Rancher Program, which includes educational programs for women and minorities as specific program priorities. Even so, at a broader level, the USDA has a checkered past on civil rights issues. In 1999, the Secretary of Agriculture commissioned a report on civil rights at the USDA that produced damning results of USDA's progress on civil rights. The USDA report states that the "USDA has not effectively protected, supported, or promoted small and limited-resource farmers and ranchers and other underserved customers. Not only have they often not been served at all, but in many cases the service has appeared to be detrimental to the survival of minority and limited-resource farmers" (USDA 1997, 30). In 2012, the USDA offered a $1.3 billion discrimination settlement to women and Hispanic farmers.[5]

WHAT HELPS TO EXPLAIN COOPERATIVE EXTENSION'S LIMITED ATTENTION TO WOMEN FARMERS?

We identify four issues that we believe contribute to Cooperative Extension's limited response to the growth and needs of women farmers. The first of these barriers is stereotypes of women prevalent in our society as a whole, which suggests that women working in male-dominated occupations are not as competent or committed to the work, and should, instead, focus on their roles related to family and home. The experiences of women farmers trying to access education and training through Cooperative Extension mirror the experiences of those in the nonagricultural labor force. Research on other traditionally male-dominated occupations has shown that when women managers or trainers hold more traditional stereotypes and values, women are less likely to participate in the labor force and are less likely to participate in outside of work or on-the-job training (Antecol and Kuhn 2000; Fernandez 2007; Kosteas 2013).

A second barrier is the stereotypical definition of a farm, which limits the extent to which women are perceived as real farmers by institutions such as Cooperative Extension. In focus groups, interviews, and surveys among Extension personnel, we found that Extension educators and administrators had conflicting viewpoints about whether women farmers were "real" farmers because of the scale and type of farms they operate (Brasier et al. 2009; Trauger et al. 2010). In general, women's farms reflect a diversity of crops and livestock, necessitating a response from multiple academic specialists at the university and the research related to the control of pests and diseases caused by the interaction inherent in the diversity of crops and of livestock (Kremen and Miles 2012). The farms tend to be unlike more conventional and economically dominant commodity farms, where a homogeneous response from one or, perhaps, two academic areas can provide the necessary educational training and research-based information. As public funding for agriculture continues to shrink, the land-grant system tends to benefit large-scale farms, while small-scale and diversified farms remain underserved (Ostrom and Jackson-Smith 2005).[6]

A third reason for Cooperative Extension's limited response is the type of educational background educators bring to the position and receive on the job. Educators are typically trained in an expert-delivery model, which does not encourage participation of the audience, a known value that women farmers express. They often are content experts with limited backgrounds in adult education or pedagogical techniques. Extension educators typically receive training on civil rights to ensure that all people have access to Cooperative Extension's educational programs regardless of race, gender, ethnicity, and sexual orientation; some may, as described earlier, conflate equal treatment with equal opportunity.

A fourth reason for Cooperative Extension's limited response is its focus on technology transfer as a measure of success, including complete technology packages (such as herbicide tolerant seeds, herbicides, and tillage practices that work together). Dramatic changes in programming strategies and communications technologies used by the Cooperative Extension organization and an Extension bureaucracy that focuses on standardized programming to support the agricultural industry may have limited the ability of individual Extension educators to address the educational needs of a changing clientele. Meeting the information and educational needs of the growing diversity of people engaged in agriculture, with a broad range of experience and agricultural education, and unconventional approaches to farming and marketing requires a very different model than the top-down, expert-to-learner technology transfer model. This model, which stresses uniformity, may still be relevant for industrialized production of major commodities but is likely inadequate to meet the needs of those who operate small to midsize highly diversified farms with complex production and marketing systems. These types of alternative systems include many women farmers.

Nongovernmental Farm Organizations

In addition to the public support for agriculture embodied in Cooperative Extension, an array of nongovernmental organizations serves the interests of farmers in the US. Here we discuss women's involvement in general farm organizations, commodity organizations, and sustainable agriculture organizations. We highlight the limited roles for women in these organizations, particularly in realms that were not consistent with traditional divisions of labor, gender ideologies, or with views of women as producers.

GENERAL FARM ORGANIZATIONS

The most long-standing, prominent, and influential organizations are broad-based farm organizations (Browne 2001). These organizations represent agriculture broadly, such as the National Grange, which was created in 1867; the National Farmers Union, founded in 1902; and the American Farm Bureau, which began operating nationally in 1919. These organizations represent all types of farmers, focus on economic and social issues for farmers, and lobby for increasing government services to agriculture (Browne 2001). Early successes of these general farm organizations included lobbying for general improvements for farmers, including lower rail transportation costs, regulation of grain warehouses, rural electrification, and price supports for farmers. These general farm organizations continue to support farmers and advocate for agricultural policies. While women have been involved at some level in all of the organizations, women's leadership roles and the focus on issues specific to gender remain uneven.

The National Grange proved exceptional in its encouragement of women's participation. Caroline Hall, who was a longtime secretary to the founder of the Grange, insisted that women be involved as members on an equal footing as men. She has been credited with convincing the Grange leadership of the importance of including women as members and was later named an honorary founder of the Grange. The American Farm Bureau has a separate Women's Leadership Committee, which gives women the opportunity to be involved in the Farm Bureau and empowers women to speak publicly about agriculture, agricultural policy, and legislative issues. Their theme for 2014–2015, "Growing Strong," emphasizes traditional roles for women such as advocating for agriculture among consumers at the state and national level. They also describe efforts to teach women business planning, social media techniques, and skills to develop local conversations about food and the importance of agriculture. The theme for 2012–2013, "Engaged, Empowered, and Strong," emphasized women's leadership within the Farm Bureau and women's importance in providing education about agriculture to the general public.[7] The National Farmers Union promotes the survival and economic well-being of family farms. They sponsored a National Farmers

Union Women's Conference in 2014 to encourage women to enhance their knowledge of their family farm operation and learn leadership skills.[8] But the gender composition of their board of directors in 2014 included one woman out of twenty-five directors.

These organizations have historically been male-dominated, although women have participated as members. While women now participate in all of these organizations, none of these organizations have dealt explicitly with confronting gender inequality on farms or in the agricultural community more broadly. However, recent efforts by several of these organizations have resulted in programs that attempt to emphasize women's importance to family farms and to build women's leadership skills, particularly in relation to advocacy for agriculture and the issues each organization promotes.

COMMODITY ORGANIZATIONS

The array of commodity organizations is astonishing, with multiple groups representing single commodities. The most powerful commodity groups represent growers of major commodities with the greatest economic interests, including cotton, corn, sugar, wheat, soybeans, cattle, dairy, and hogs (Browne 2001). As discussed in previous chapters, women are least likely to be principal operators on farms and ranches that raise these commodities. Commodity organizations typically lobby for improved economic benefits for their producers, including price supports, market access, and disaster relief. In addition to these powerful major commodity groups, other organizations represent growers of every type of farm product, ranging from the North American Raspberry and Blackberry Association, the National Onion Association, to the American Goat Society.

Women's involvement in commodity organizations was initially accomplished through the establishment of women's auxiliaries beginning in the 1960s. For example, the Iowa Porkettes was established in 1964 as an auxiliary of the Iowa Pork Producers' Association. Their primary goal—very much in line with appropriate roles for women at the time related to food and consumption (and not production)—was to promote pork consumption through promotional events at fairs and supermarkets, conducting annual Pork Queen contests, and introducing information about pork into school curricula. Their archives reveal that as they grew in size and professionalism, some members raised concerns that the name "Porkettes" was diminutive and unprofessional and that the Pork Queen did not reflect women's role in pork production. A vote in 1984 resulted in keeping the name "Porkettes" and the Pork Queen because the position had developed into a professionalized spokesperson for the pork industry. Porkette members gradually became members of committees of the Iowa Pork Producers Association, and in 1991 the Iowa Porkettes disbanded.[9]

Like the Porkettes, women's auxiliaries of commodity organizations have largely disappeared as women have stepped into the main membership of the organizations. However, some auxiliaries still remain, and tend to focus on women's nonproduction roles (such as food, consumption, and education). American National CattleWomen for example, conducts programs specifically designed to promote beef consumption to consumers and include educational programs about nutritional qualities of beef for K–12 school programs, cook-offs, and beef recipes. In addition to their national organization, various states such as Florida, Utah, and Ohio have state-level organizations of cattlewomen. Some of these organizations are beginning to confront their portrayal of the role of women in the beef industry. For example, in 2013, the Ohio Cattlewomen's Association retired the Beef Queen. The trend now is that instead of having gender-segregated auxiliaries, women participate directly in the commodity associations. However, for the most part, these organizations have not adopted policies to examine gender issues in an effort to improve gender equality. These auxiliaries focus more on women promoting food and consumption rather than their roles in agricultural production.

SUSTAINABLE AGRICULTURE ORGANIZATIONS

The emergence of sustainable agriculture over the last few decades has also resulted in the growth of organizations to serve these farmers. Sustainable agriculture organizations provide a policy voice for small- and medium-sized farms that often use organic or sustainable production practices, are diversified, and produce for local markets. Many of these organizations work at state or regional levels such as the Pennsylvania Association for Sustainable Agriculture, Practical Farmers of Iowa, the Land Stewardship Project, and the Midwest Organic and Sustainable Education Service. These organizations provide educational programs for farmers, encourage farmer-to-farmer exchange of information, and advocate for state and federal policies to promote sustainable agriculture. Sustainable agriculture organizations arose to support operators of small- and medium-sized farms who were not having their educational needs met by Cooperative Extension and not having their policy interests served by general farm or commodity organizations. Compared to other farm organizations, women are more involved as members and, to some extent, as leaders. Women farmers are more likely to be involved in sustainable agriculture than in conventional agriculture (Trauger et al. 2008). Compared to other farm organizations, women are more likely to be on boards of directors. For example, in 2012, women were 36% of board members of Practical Farmers of Iowa, 44% of board members of the Pennsylvania Association of Sustainable Agriculture, and 60% of board members of the Land Stewardship Project. Despite the higher level of participation of women in leadership positions, these organizations rarely advance gender equity on farms or in agriculture as one of their primary issues.

Women Form Their Own Agricultural Organizations

In the mid-1970s, some farming women formed their own organizations in response to unstable conditions in agriculture, including declines in commodity prices and rising fuel prices (Devine 2013). These organizations differed from male-dominated farm organizations in which women played supporting or auxiliary roles. Two of the most prominent organizations at the time were Women in Farm Economics (WIFE) and American Agri-Women. WIFE was formed in 1976 by a group of farm women who met in Sidney, Nebraska, to discuss problems in agriculture and to advocate for agriculture. Devine's analysis of WIFE explains how the women in WIFE "carefully navigated the masculine worlds of agriculture and politics and sought to refine a unique, public voice for farm women in politics" (Devine 2013). Devine explains that the women of WIFE used organizational strategies drawn from the civil rights and second-wave feminist movements that promoted female leadership; however, they did not embrace women's equality and even emphasized their dependence on men, as reflected in their name. Their emphasis was on farming in general, not on women in farming. They lobbied on behalf of farming, spoke on radio and television shows, and publicly advocated for the well-being of agriculture, while always positioning themselves as dependent on their husbands.

American Agri-Women formed in 1974 as a coalition of farm women's organizations in various states to advocate for agriculture and address the problems facing it. The first of these organizations started when a group of women in Oregon decided they needed to effect change in government policies to protect farmers because their husbands were too busy farming to respond to government policies. From here, these Oregon women met with Women for the Survival of Agriculture in Michigan and decided to hold a national conference, which resulted in the coalition of multiple state organizations that formed American Agri-Women. The group describes itself as the largest coalition of farm, ranch, and agribusiness women in the US with the goals of informing consumers about agriculture, lobbying congress for farm legislation, and supporting policies promoting large-scale, industrial agriculture.[10] They rarely bring to the forefront women's issues in their programs, although they do encourage women to step into leadership positions in agriculture.

Summary: Women and Traditional Agricultural Organizations

While empowering each other to speak and provide a certain type of leadership in public spaces, most farm women's organizations were careful not to threaten male leadership or call too much attention to the patriarchal

Women Farmers' Reasons for Belonging to Farmer Organizations (%) (SNEWF)

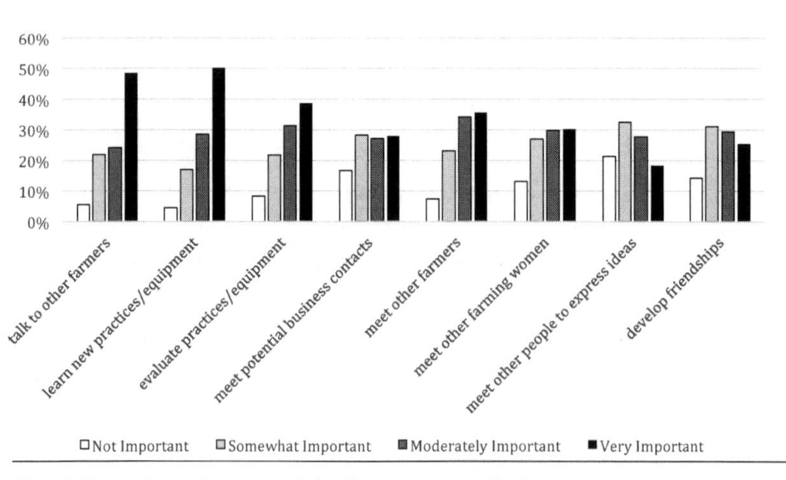

Figure 4. Women farmers' reasons for belonging to farmer organizations.

2 NORTHEAST WOMEN FARMERS AND ORGANIZATIONAL MEMBERSHIP (SNEWF)

Our survey of women farmers in the Northeast included questions about organizational membership. Of the 815 respondents, 678 (83%) stated that they belonged to or had participated in at least one farm-related organization within the three years. Of those who belong to any organization, about one-third (36%) belong to one organization, 41% belong to two organizations, and 23% belong to three or more organizations. The most popular type of organization is sustainable or organic agriculture groups (51% indicating membership in the past three years), followed closely by general farm organizations (48%), commodity organizations (34%), and farm women's organizations (19%).

The primary reasons why women farmers in the Northeast join these organizations are a mix of education, skill-building, and personal and business network development, and are consistent with the discussion in this chapter. Specific reasons include learning new practices/equipment (79% state as "very important" or "moderately important"), talk to other farmers (72%), evaluate practices/equipment (70%), meet other farmers (70%), meet other farming women (60%), meet potential business contacts (56%), develop friendships (54%), and meet others to express ideas (46%). Advocacy is also important, with over half of all women stating that advocacy for changes to the food system (71%), for sustainable farming (71%), for farmers (67%), or for women farmers (57%) is a moderately or very important reason for joining agricultural organizations.

Figure 4 shows a further breakdown of these percentages based on the levels of importance and lists the percentages for "very important" and "moderately important" separately.

organization of family farms. They worked on agricultural issues such as commodity prices and government agricultural policy rather than issues of gender inequality (Devine 2013). Acting in this way maintained their status in the existing patriarchal system but did not create space and opportunities for women in agriculture who did not fit into this system. For those women who strove to see themselves, and have others see them, as producers, these traditional agricultural organizations created barriers to their success. Most organizations have identified women in agriculture only through their relations to men (as farm wives) and in reproductive roles (as homemakers).

Agricultural organizations can provide valuable technical, financial, and business resources that are important for growing farm businesses. For women, this need may be even greater; as noted in earlier chapters, women tend to come to farming with less background, and consequently need the assistance to create successful enterprises. Further, they also need social support as women in a male-dominated occupation. In response, women farmers have begun to create organizations that fulfill these needs, to which we turn next. These new network-based organizations serve women who are farmers (either alone or in partnership with others), support the full participation of women in the agricultural system, and provide opportunities to learn, support, and network with each other.

Creating New Networks to Meet Women Farmers' Needs

In response to the lack of adequate organizational support, women farmers have created agricultural networks that are intentionally focused on providing access to the resources they need to successfully overcome the challenges all farmers face, as well as the challenges women specifically face. These new networks of women farmers differ from the agricultural organizations described earlier because of their direct emphasis on including the voices of women farmers to establish the networks' agendas, to promote gender equality in agriculture, and to gain recognition of their members as producers. These networks offer knowledge- and skill-building opportunities and a network of women with ideas that have been tested on their own farms. The networks also offer social support that can alleviate isolation, legitimate women's identities as farmers, and increase their capacity as farmers, ultimately enhancing the success of women's farm businesses. One member of PA-WAgN described the impact of participating in the network by contrasting the self-doubt she feels when away from the network to the sense of affirmation she feels when she participates in PA-WAgN events:

> When I am at home, sometimes I feel like that saying "fake it till you make it." People put me in this space—"Are you really a farmer?" Yes I am. I feel empowered here [in the network]. I feel recognized for who I

> am and what I am doing when I am with other women farmers, and that gives me strength to go home and keep doing what I want to do.

This quote illustrates the value of women famers' agricultural networks and the essential role they can play in supporting women as farmers and in providing a collective voice for women farmers.

In this section we first describe the history of women's agricultural networks (WAgNs) in the US to illustrate their growth and emphasis on women farmers' needs. The Vermont Women's Agricultural Network, Maine Women's Agricultural Network, and Iowa's Women, Food, and Agriculture Network were the first women's agricultural networks in the United States. We then describe the Pennsylvania Women's Agricultural Network (PA-WAgN) to illustrate how a network can function, the strategies used to identify and then meet the needs of women farmers, and finally, the impact of the educational content and social support. We describe how PA-WAgN has helped spark women farmers' empowerment through a network that provides technical training and builds social capital among women farmers. This empowerment has led to multiple outcomes, including a greater sense of identity as a farmer, greater independence and sense of legitimacy on the farm, the search and discovery of agricultural alternatives that fit the goals of women farmers, more innovative and successful farm businesses, and greater engagement in community and agricultural efforts.

Based on our research and experiences with PA-WAgN, we find that the growth of a women's agricultural network offers further empirical evidence of a transition to a feminist agrifood system in which some women farmers confront the gendered barriers in agriculture and devise innovative ways to overcome them. The networks have become an essential aspect of developing gender equity in agriculture for the women who have access to them. Additionally, the networks have become resources to help women secure their place in one of the primary industries of our society: agriculture.

Pioneering Women's Agricultural Networks

Here we briefly profile three groups that laid the foundation for the development of a new type of agricultural organization for women farmers. These founders developed network-based organizations that emphasized education and social support for women farmers.

VERMONT WOMEN'S AGRICULTURAL NETWORK

Mary Peabody, a community and economic development specialist with the University of Vermont Extension Service, worked with small-scale farmers, many of them dairy farmers transitioning out of dairy farming at a time of an oversupply of milk (Peabody 2012). At Extension events, she noticed that women attending the meetings would not ask questions during the regular educational sessions but mobbed her at the coffee break. She thought that women were intimidated about speaking up in meetings attended primarily by men, and after deciding that women needed women-only meetings, she obtained support from one of the leaders of Cooperative Extension in Vermont and a six-month planning grant from the USDA to develop a program for farm women. Mary and her team designed a program based on discussions with women farmers, aspiring farmers, women who had left farming, and other relevant stakeholders, such as financial institutions and others who could provide resources to women farmers. Based on these discussions, Mary and her team formed the first Women's Agricultural Network in 1995. A five-year grant from the USDA provided funding stability to establish a program focused on educational and technical assistance in farm business planning.

Mary Peabody differentiated Vermont WAgN from other women's agricultural organizations based on the role they assumed women have on the farm. As she described, other women's agricultural organizations defined farm women primarily as homemakers and targeted farm wives as opposed to women who were primary decision-makers and farm operators. By contrast, Vermont WAgN worked with women who identified themselves as farmers, decision-makers, and leaders. Vermont WAgN did not explicitly advertise as a feminist organization. Mary explains, "Our way of being in the world was feminist—inclusive—asking people what they wanted, offering interactive education, and providing childcare." Vermont WAgN worked mainly with farmers involved in sustainable agriculture, and although it reached out to conventional networks, it found that these women tended to identify themselves as farm wives rather than as farmers. Clearly, this new type of women's farming organization was developing out of the needs of women farmers who were occupying new roles.

MAINE WOMEN'S AGRICULTURAL NETWORK

Vermont WAgN served as a model for other women's agricultural networks. Based on Mary Peabody's efforts in Vermont, Vivianne Holmes, an agricultural educator from Maine Cooperative Extension, spearheaded the establishment of the Maine Women's Agricultural Network in 1997; by 2004, Maine WAgN provided support to almost one thousand women farmers in Maine and New England (White 2004). Maine WAgN had a diverse set of supporters including the University of Maine's Cooperative Extension, the Maine Organic Farmers and Gardeners Association (MOFGA), the Heifer Project International, USDA's Natural Resources Conservation Service, the Western Mountains Alliance, and individual women farmers. Maine WAgN implemented a variety of participatory activities, which included farm tours by women who traveled around the state to visit other women's farms, workshops on topics such as chain saw safety conducted by regional subgroups of women farmers, and informal workdays for women to help each other on their farms. One of the subgroups included the Daughters of Yarrow, a group of lesbian farmers who met on each other's farms to help with work and share their experiences (White 2004). Maine WAgN is unique among the women's networks in providing a separate space for lesbian farmers to network. Although Maine WAgN disassociated from Cooperative Extension when Vivianne retired, its current Facebook page indicates that it remains "a collaborative effort of farmers, agriculture, and related non-profit, for-profit and governmental agencies seeking to help women to develop the skills necessary to begin or continue farming and managing woodland profitably."

IOWA'S WOMEN, FOOD, AND AGRICULTURE NETWORK (WFAN)

As Vermont and Maine led the new wave of women farmer networks in the Northeast, the Women, Food, and Agriculture Network (WFAN) initiated support for midwestern US women farmers in Iowa in 1997. The Iowa-based network stemmed from the work of Denise O'Brien, a woman farmer and agricultural activist from Iowa, and Kathy Lawrence, a sustainable agriculture activist from New York, to remedy the absence of women's voices on food and agricultural issues prior to the United Nation's Fourth World Women's Conference in Beijing, China, which established a Women, Food, and Agriculture working group in 1994. After Beijing, Denise formed the network to focus on women's issues related to rural, agricultural, and environmental problems.

WFAN differs significantly from the northeastern US women's networks. It is not part of a land-grant university or Cooperative Extension, although it works informally with faculty and students at Iowa State University. Based in Iowa, with most members from Iowa or neighboring states, WFAN has reached out geographically, and members now represent twenty-five states and other countries. The organizations communicate with one another in a loose network.

Unlike Vermont and Maine WAgN's emphasis on education, WFAN has been more involved in advocacy for women's leadership in agriculture. WFAN organizes annual meetings for women involved in sustainable agriculture, advocates for women farmers, encourages women to serve on agriculture-related boards and commissions, and assists women landowners with their decision-making about land use. One signature WFAN program, Women, Land and Legacy, focuses on women who own farmland but who do not necessarily operate a farm. A follow-up program, Women Caring for the Land, motivates women landowners to increase their efforts in conservation and sustainable agriculture (Carter 2014). Although WFAN does not use WAgN terminology, its focus on serving women as farmers and decision-makers is similar to the WAgNs of the Northeast. The pioneers in these organizations have led the way for other states to form women's agricultural networks, including Pennsylvania, and signal the beginning of a movement toward organizations that recognize women as producers and leaders in agrifood systems. Next we provide in-depth information about the Pennsylvania Women's Agricultural Network (PA-WAgN) as an illustration of how these networks work and their potential impact on women farmers.

The Pennsylvania Women's Agricultural Network (PA-WAgN)

The Pennsylvania Women's Agricultural Network (PA-WAgN) began in 2003 in response to the needs and circumstances of women farmers in the Mid-Atlantic region. Box 3 provides a firsthand account from one of the founders of the formation of PA-WAgN. This narrative identifies a number of key issues and principles that have formed the basis of the ways in which the network is structured and how it operates, including a focus on participatory action research, negotiating relevance within patriarchal institutions, using collaborative education for empowerment, and focusing on alternatives to conventional agriculture. These themes are explored in greater detail below.

The founders and several members are associated with a land-grant university and use feminist and participatory action research that emphasizes collective inquiry, experimentation grounded in experience, and social history to inform the structure and programming of PA-WAgN (Kindon et al. 2007). In early research by the university-based founders, women farmers revealed their sense of isolation, unwelcoming atmospheres in other agricultural groups, and women farmers' lack of access to resources and information. It is these essential findings that have driven the goals and programming of the network.

The institutional connection with the university provides opportunities and critical resources to demonstrate the value of a feminist approach but also has created challenges. This connection has resulted in an organization that has emphasized research and educational programs and has drawn from the institution's resources to create the network. Trying to form a feminist organization within the confines of an organization that does not operate on feminist-inclusive principles has resulted in a persistent tension with the university administrative and Extension systems for resources and recognition. While the university provides faculty time, office space, and access to communication technology to PA-WAgN, all operating funds and support for staff are derived from extramural, grant-based funding. The relationship to the university has been a topic of conversation since the founding and continues to arise regularly.

PA-WAgN's educational programming is guided by research and ongoing assessments through which women describe their educational needs. The educational programs are designed in response and reflect a holistic vision of education that extends beyond technology transfer. It includes an awareness of the social embeddedness of educational settings, the information being shared, and the people who are participating in the activity. A core principle guiding PA-WAgN structure, planning, and events is empowerment of all participants. The network intentionally designs educational activities to

3 FORMATION OF THE PENNSYLVANIA WOMEN'S AGRICULTURAL NETWORK

A small group of women farmers and women agricultural professionals started PA-WAgN in 2003 with the goal of providing educational and networking opportunities for women farmers. The inspiration for PA-WAgN came from a presentation on the accomplishments of Maine's Women's Agriculture Network at the Rural Women's Studies Association meeting in New Mexico.

Both Carolyn Sachs and Amy Trauger had been conducting research with women farmers in Pennsylvania and other parts of the US and the world. Amy's research in Pennsylvania found that women farmers experienced isolation as they tried to build their farm enterprises. At the New Mexico meeting, Carolyn and Amy met with two women from Maine's network and decided to pursue the idea of a similar network for women farmers in Pennsylvania. Upon returning to Pennsylvania, they organized several meetings of women farmers, agricultural educators, and other agricultural professionals to discuss the possibility of starting such a network in Pennsylvania. Of course, people shared different ideas of how the network should operate, who should be included, and what types of activities the network should engage in. The group decided to seek advice from the wisdom and experience of other women farmer networks.

The Pennsylvania founders invited the leaders from the Vermont Women's Agricultural Network, Mary Peabody, and the Maine Women's Agricultural Network, Vivianne Holmes, to Pennsylvania to meet with the developing group. Everyone in the group worked in agriculture in some way and included farmers, students, faculty, and staff in the college of agriculture at the local university, representatives of farmers' organizations, and state department of agriculture employees. All had one thing in common—a sense of dissatisfaction with the lack

provide members an inclusive and welcoming space for innovation, where creativity is fostered and celebrated. PA-WAgN events are structures to reinforce that everyone has knowledge to share and can be both teacher and learner in a given situation. The emphasis on peer education, mentoring, networking, and interactive events are hallmarks of PA-WAgN and result in women's empowerment and sense of legitimacy as farmers. In the following sections, we further describe the structure and ways of operating within PA-WAgN that reflect these principles. We then turn to the impacts of this approach on PA-WAgN members.

of recognition, educational programs, and institutional support for women in agriculture on farms and in other agriculturally related institutions—so they set out to shape a new model for women in agriculture to work together.

The event was held at Haley Selkirk's farm in the midst of a dairy farming valley in central Pennsylvania. Haley runs a farm camp, so they settled into a meeting room with a view of the rolling hills, mountains, and cows; slept in rustic unheated cabins on a cold October night; and savored delicious meals. While some of the women knew each other, many were meeting for the first time. The group energy was infectious—ideas, visions, and connections abounded. They quickly learned that they held different ideas depending on the types of farms they operated and which organizations they worked with, different levels of trust in and affiliations with traditional agricultural research and Cooperative Extension institutions, and different levels of energy to put into a new organization.

With the guidance of Vivianne and Mary, they developed a mission statement and officially started the Pennsylvania Women's Agricultural Network. They discussed who should be included in the organization, if they should be a membership organization, how to span the entire state, what would be their primary activities, if they should focus on sustainable agriculture or reach out to women who are farming on conventional farms, and what would be the role of the university in the organization. After extended discussion and debate, they arrived at the following mission statement: "The Pennsylvania Women's Agricultural Network (PA-WAgN) supports women farmers and women agricultural professionals by providing positive learning environments, networking, and empowerment opportunities."

The Structure of the Pennsylvania Women's Agricultural Network

The organizational structure of PA-WAgN draws on and links two strengths: (1) a network of women farmers and agricultural professionals that provides experience, ideas, and social support, and (2) a land-grant university and Cooperative Extension system to conduct research and develop educational resources. These resources are drawn together through the ways in which the four key parts of the organizational structure—PA-WAgN members, a steering committee, a working group, and a research group—work together to provide both long-term guidance and day-to-day management of the network (fig. 5).

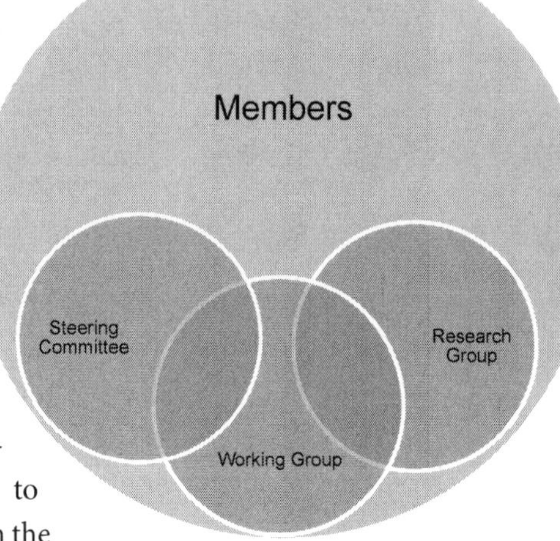

Figure 5. Organizational structure of PA-WAgN.

MEMBERS

Membership in PA-WAgN grew from about 100 members in 2004 to 1,611 members in 2014 (fig. 6).[11] Membership is free, and members can join at various events or online. Free membership reduces barriers of participation for people with limited resources. Most women farmers have a level of socioeconomic status that enables them to farm and allows them to participate in the network. However, the costs of attending events might be burdensome to some women due to time, travel, and distance. PA-WAgN uses multiple communication tools (its website, Facebook, Twitter, mailings, online and printed newsletters, email announcements) to provide information about PA-WAgN and its events. Members are not just passive recipients of information; they teach each other, act as mentors, plan events, and lead educational and networking events at their farms, businesses, and elsewhere. They regularly provide guidance about topics for field days and direction for PA-WAgN.

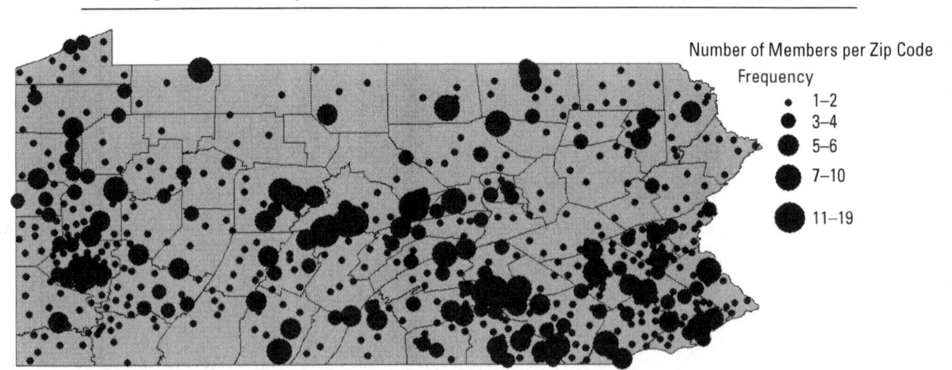

Figure 6. PA-WAgN membership as of December 2014 (members outside of Pennsylvania not shown).

THE STEERING COMMITTEE

The steering committee, composed primarily of women farmers, sets the overall goals for the network, helps to identify needed educational programs and the resources to conduct them, and draws on their individual and organizational resources to further the leadership and mentoring goals of the network. The farmer members of the steering committee represent their region, serve as mentors to farmers in their regions, and provide an important point of contact and connection between PA-WAgN and farmers in their region. The regional representatives help plan field days, potlucks, and other events relevant to their area. Other members of the steering committee represent organizations relevant to women farmers in Pennsylvania, including government agencies (e.g., USDA's Farm Service Agency), educational institutions (e.g., Penn State Extension, Delaware Valley University, Chatham University), and agricultural nonprofit organizations (e.g., the Pennsylvania Association for Sustainable Agriculture, the Rodale Institute). Since 2004, the steering committee has met four times per year, comprised of two face-to-face meetings and two conference calls, to plan future activities, connect across regions, and identify funding opportunities.

THE WORKING GROUP

The working group at Penn State University is composed of a small number of faculty, staff, and graduate students. They provide the day-to-day administration of the network, which includes maintaining communication, managing budgets, organizing steering committee activities, and collaborating with members to plan and organize events. Currently two full-time staff are supported through extramural funding.

THE RESEARCH GROUP

The research group, composed of a small number of faculty and graduate students, collaboratively conducts research with PA-WAgN members and other women farmers and interprets and disseminates research results. The research produced by this group includes needs assessments and evaluations of educational programs, which in turn informs network activities. The research also contributes to scholarly knowledge and conversations on women farmers and agrifood systems and the publication of scholarly articles. The group also collaborates with the working group to write grant proposals to support the staffing and programming of the network. This research and interaction with members has culminated in the writing of this book.[12]

PA-WAgN Partners

PA-WAgN engages with many different partners, including Cooperative Extension, sustainable and organic agricultural organizations, and federal, state, and local government agencies. PA-WAgN has collaborated with specific educators in Cooperative Extension to offer educational programs and leadership training. For example, Extension educators have participated in or led field days on beekeeping, organic soil management, high tunnel production, and sheep production. It partnered with several educators to revise Annie's Project (described earlier in this chapter). Extension educators also led leadership training for the PA-WAgN steering committee members. Partnerships with sustainable agricultural organizations include the Pennsylvania Association for Sustainable Agriculture and Pennsylvania Certified Organic. PA-WAgN regularly conducts hands-on, preconference workshops at the annual conference of the Pennsylvania Association for Sustainable Agriculture on topics of interest to women farmers, such as using and maintaining farm equipment and building farm infrastructure. It has also partnered with the USDA's Natural Resources Conservation Service to offer field days on conservation and farmer access to government programs. PA-WAgN also partnered with a community development organization in central Pennsylvania to develop a farmers' market in an area designated as a food desert (see box 4). PA-WAgN also works with other

---

4 COOKSHOP

PA-WAgN organized a weekly cookshop located near the market that involved local residents and farmers teaching people how to cook and use produce from the farmers' market. At the farmers' market, PA-WAgN sponsored a ten-week class series that featured nutritionists, chefs, farmers, and skilled home cooks who provided demonstrations on how to make easy homemade baby food, how to make soup from scratch using leftovers, creating fresh salsa, saving time with slow cookers, ten ways to eat a zucchini, and cooking with kids. Partnerships such as these have enabled PA-WAgN to reach a broad audience.

Maryann Frazier demonstrating a healthy hive during a PA-WAgN hands-on apiary workshop. Photograph by Ann Stone, PA-WAgN.

women's agricultural networks to organize conferences and seek funding for its efforts.

PA-WAgN Events: Education, Networking, and Leadership

The voices of women who have participated in PA-WAgN research and who interact with other members led us to recognize the value of, and need for, events that focus on three intertwined components: technical farm-based and peer-to-peer education, networking to build social support and business connections, and opportunities for enhancing leadership skills. In this section we describe PA-WAgN events that address all three.

Between 2005 and 2014, 3,542 people attended 125 educational events organized by PA-WAgN. Field days on women's farms constitute the majority of events, and have covered a wide range of topics that include operating farm equipment, marketing, business planning, producing small ruminants, producing vegetables and fruit, and adding value to production. Conservation, sustainable, and organic production include several other topics. Events are typically hosted and led by farmers with support from technical experts as needed. All events have included hands-on learning and time for networking. The objective of these events has been to build not only

5 PA-WAGN FIELD DAY SUMMARY: SLAUGHTERING CHICKENS (AS RECOUNTED BY ONE PARTICIPANT)

Dara, her husband, Kevin, and Sarah, her daughter, conducted a field day on chicken processing on their diversified animal and vegetable operation. The event was held in their new farm building, which was designed and built by Kevin to serve multiple purposes of processing chicken, processing fruits and vegetables in a commercial kitchen, serving meals to customers, and selling directly to customers at their farm.

The day began in the meeting and dining area with Sarah and a certified commercial kitchen inspector reviewing food safety in chicken processing. We washed our hands, then took turns using a special dye that showed how well we had removed the bacteria from our hands. Those of us who tended to wash our hands much too quickly or skimp on the soap found our hands marked bright purple. This "hands-on" experience made an indelible impression! After properly washing our hands, we moved to a small outdoor area adjoining the building to begin slaughtering and processing chickens.

PA-WAgN limited the number of attendees to provide each participant with the opportunity to process their own chicken. Kevin led us carefully through the step-by-step process of killing chickens and proudly showed us their low technology but very effective defeathering machine. Then Sarah demonstrated how to dress the birds. Each field day participant had the opportunity to dress a bird and most of the women attending jumped right in. Sarah and Kevin patiently demonstrated how to clean the birds regardless of how squeamish some people felt as they watched their first slaughter. Everyone needed extensive help and asked an endless number of questions. After putting the chickens in the freezer and washing up, we returned to our meeting room, which had been transformed into a dining room, and sat down to a fabulous lunch (not chicken). Dara gave us a tour of the entire operation, including the vegetable fields, the hoop house, and the barns. By late afternoon, everyone was tired and headed for home. Everyone in attendance profusely thanked Dara, Kevin, and Sarah for a fantastic workshop, and many mentioned they were inspired to enhance or start small-scale chicken processing on their farms. Who ever imagined slaughtering chickens could be so inspiring?

technical knowledge about running a farm and farm business, but also to build social and business relationships among women farmers. Two examples of field days illustrate the elements of these events (see box 5 and 6).

In addition to the educational field days, PA-WAgN holds an annual all-day network-wide conference as an educational and networking event. These

6 PA-WAGN FIELD DAY SUMMARY: MARKETING VEGGIES AND SAUSAGE

Another PA-WAgN field day featured Robin and Kent's farm, a diversified vegetable, grain, and beef operation. On a sweltering July day, we all crowded under the shade of a carport to learn about high tunnel vegetable production. High tunnels are simple hoop buildings that protect crops and extend the growing season. Kent described in vivid detail the ins and outs of high tunnel production and answered questions about pest control, water management, varieties, and more. We also learned that the carport we were huddled under served as their farm market stand until they opened their new farm store a month before.

Kent emphasized how he and Robin were equal partners in their operation. Robin, seated in her wheelchair, added her insights and corrected him as she deemed appropriate. He said that you might not think they were equal partners because she is confined to a wheelchair, but he explained how they had the truck retrofitted so that she could use it without his help, and he also proudly described how she drove skid loaders in the barn.

We then drove several miles to their new farm store to learn about their marketing strategies and also their partnerships with Hank, a veterinarian who processes and markets sausages. Hank, Robin, and Kent all described the process of building a commercial kitchen that met the standards for meat processing including the cooling system, appliances, water testing, and helpful hints of how to deal with state inspectors.

We tried Kent's different sausages—hot, mild, burgundy, or vegetarian (made especially for the field day). Robin's sister and niece and Jay's sister and brother-in-law prepared the delicious lunch from the high-tunnel vegetables. The meal was topped off with the pastries from their kitchen, which are also sold at their market.

After answering a slew of questions and entertaining us all, Kent and Robin exclaimed what a pleasure it was to have this educational event at their farm. Kent said they never realized how much information they had to share with other farmers.

events involve keynote speakers, such as Karen Washington, founder of Black Urban Growers; Judy Wicks, founder of the White Dog Cafe and Fair Food of Philadelphia; farmer presenters; representatives from not-for-profit organizations; and Extension educators. The number of participants ranges from 100 to 145, with many new and beginning farmers attending the event. Topics range widely and depend on what members have expressed as a special need or interest. For example, a recent conference included workshops on production and management issues that included how to operate a high tunnel, how to work with interns, and how to farm with a nonfarming partner.

Steering committee members and regional representatives often organize informal events at their farms, such as networking potlucks. Women farmers meet at each other's farms, bring food, and share their experiences. Sometimes the organizers focus on specific work tasks that they need help with on their farm, which becomes an opportunity to learn more about farming, such as chain-saw use and safety and building a hoop house. At other times, the host farmers offer to share information about a topic in which they are specifically interested, for example, requirements for certified kitchens or how to write grants. We include an example of a typical PA-WAgN potluck gathering in the box 7.

Finally, the annual winter meeting of the steering committee includes leadership training in addition to PA-WAgN business. Topics of past work-

---

7 POTLUCK EXAMPLE

Ten women gathered at Mary and Roberta's farm to share food and learn how to safely operate chain saws. One of the participants, Joy, who had extensive experience as a logger at a sawmill operation, carefully led the women through chain saw safety, maintenance, and use. Many of the women brought their own chain saws, which they used either reluctantly or without much training. Starting the chain saw can be physically challenging, so Joy showed certain techniques for how people with less strength could fairly easily start their saws. Everyone had a chance to start a chain saw and to practice cutting a tree. The farmers benefitted because they were able to remove three trees that needed to be cut down, and they also had a new supply of firewood. At the end of the day, everyone was tired but sat down to a wonderful meal where talk and stories were shared.

---

shops have included strategic planning, how to lead difficult groups, and building women's success in organizations. For example, the 2012 winter meeting included representatives from the White House Project, a nonprofit organization that promoted women's leadership in business and politics. Liz Johnson and Catherine Grey from the project led steering committee members and PA-WAgN staff through intensive leadership training on developing personal and organizational visions. As one participant recalls, "Participants shared, explored, and took risks as we developed visions for our futures and the future of PA-WAgN. The informal interactions at meals and in the evening, where we envision major plans for international networks of women farmers, compare strategies for marketing, and update each other on our lives are the highlight of these meetings." These events create a bond among steering committee members, enhance commitment to PA-WAgN, and invest in steering committee members as agricultural leaders.

The diverse activities illustrate both the process of PA-WAgN educational events—the space for informal interaction, hands-on learning, farmer-led education, and peer-to-peer learning—as well as the content. In the next section we provide quantitative and qualitative evaluation data, which indicate that PA-WAgN events leave a lasting impression on participants, legitimizing their role as a farmer, enhancing their skills to improve their farms' success, creating personal networks, and inspiring leadership in their communities.

Impacts of PA-WAgN on Women Farmers

Multiple tools have been used to evaluate PA-WAgN events and activities, and together they indicate the strengths of PA-WAgN's impact (Kiernan et al. 2012). Across all educational events, 91% of the women farmers reported a gain in their understanding of the specific topics and skills covered in each of the educational field days and other events (see figure 7).

Impact, however, was not limited just to knowledge gains; educational events also influence how farmers manage their farm businesses. The evaluation data indicate that 77% of the women were "very" or "moderately" inspired to modify their own farm operation. PA-WAgN events also achieved two other educational objectives. The first was to empower women to search for resources beyond the WAgN event itself: 89% of the participants reported that they would seek other information and people with expertise related to their farm. The second objective was to empower

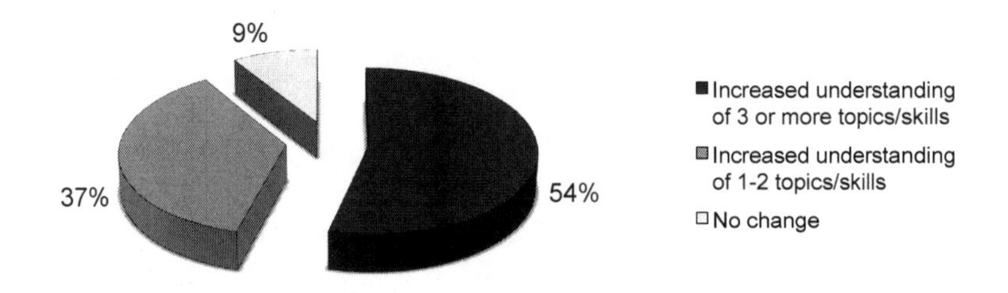

Figure 7. Percentage of women who increased their understanding of topics and skills at PA-WAgN events from 2006 to 2009 (N=303). Adapted from Kiernan et al. (2012).

women to educate other farmers: more than half the women reported they would create learning opportunities for other women farmers, indicating the importance they placed on mentoring other farmers in the network.

More than Just Education: Developing a Personal Network

A critical component of participation in PA-WAgN events has been the opportunity for members to develop a personal and business network at events (Kiernan et al. 2012). Networking is at the heart of PA-WAgN events like the field days. Evaluation data have indicated the importance of networking at events: 73.2% of attendees said they met someone with whom they will keep in contact. The attendees specified what those contacts are expected to yield for them in the way of impact: collaborate and share information (83.3%), provide technical information (50.0%), and provide business leads (36.5%). While PA-WAgN educational events have targeted the expressed needs of women farmers, these events have been available to all, and the number of men who have attended PA-WAgN events has increased over time (males comprised 16.4% of those who completed evaluations). Their attendance has most likely been due to the welcoming and nonthreatening learning environment WAgN has created for farmers and the availability of topics relevant to small, diversified farms.

The networks that women build through PA-WAgN have become critical to women farmers' success. The following sections illustrate the multiple impacts of networking for women farmers, including providing inspiration for entrepreneurial activity, mitigating isolation, providing empowerment, finding acceptance in agriculture, and using leadership skills.

BUSINESS INSPIRATION FOR FARM WOMEN ENTREPRENEURS

One way that networks can impact women's businesses is to spur innovation, particularly for farms that are struggling in traditional agriculture. The following example describes how PA-WAgN was instrumental in helping Kathleen Swift identify a business opportunity that not only provided a successful income stream but also legitimated her role as the farmer.

After farming a conventional dairy farm for twenty-two years, Kathleen Swift and her husband were in "dire financial straits" similar to other dairy farmers in their community.[13] Someone recommended to Kathleen that she get in touch with the Pennsylvania Women's Agricultural Network. She took a pen and wrote "PA-WAgN" on her hand and made contact. Leslie, a PA-WAgN member with a small farm, came to Kathleen's farm and explained that from her perspective, Kathleen was sitting on a gold mine. Leslie observed that Kathleen, with a large family farm and location near a major metropolitan market, had the opportunity to market directly to consumers.

After having connected with Leslie and PA-WAgN, Kathleen joined twenty women farmers on a road trip to a conference of the Vermont Women's Agricultural Network, where she met a woman who was pursuing an idea similar to one she had, to create a whole market-basket CSA. Since this trip, Kathleen and her family have radically changed their farm, pursuing multiple business enterprises: they converted their conventional dairy to a grass-based dairy, raised specialty grains, and began a whole market-basket CSA, while paying off their debt.

In addition to achieving success on their farm, Kathleen has changed how she sees herself, her relationship with her husband, and her input into critical decisions about the farm. Prior to her trip to Vermont, Kathleen defined herself as the farmer's wife and mother who stayed in the background of major farm decisions. Now she expresses pride in their whole market-basket CSA, and her husband gives her all the credit. As she explains, "I just love that my husband will now say it was all my idea. He says, 'I didn't want to do it, but it was totally her idea. It is working.' He says that all the time." She describes the sea change that happened with her. "I make the decisions now. We talk about it, but I basically run the show. I have experienced both validation and empowerment. I feel like I was just in the backseat to him; I was the mama who takes care of the kids, and I was just the one who made supper." Kathleen goes on to say, "[The] Pennsylvania Women's Agricultural

Network has totally changed my life—personally and professionally and our whole farm."

Other PA-WAgN members have expressed the importance of networks for sharing and evaluating business options. Maureen says that PA-WAgN provided technical information based on the real experiences of other farmers and fostered the development of relationships: "To really have in-depth discussions with people about what the costs are. That has been the most valuable thing to me, and the intimacy of WAgN and field days." Another participant says, "I know I have seen all of your farms and farms of many other women, and I get all of these ideas, and then I come back, and I try to share them with other people. It is just an unbelievable education that I would not be able to get anywhere else." Irene suggests,

> It is also just the sisterhood that is created when you are together. When I came home after our steering committee meeting, I was so pumped and charged. Just being in a room full of amazing women for a whole day, it was just, wow. I called my husband and he said, "Wow, your charisma is just oozing through the phone," and I exclaimed, "Yes, I am charged."

Dara describes how the network sparked her thinking when she was developing her own farm:

> Haley had a field day at her farm about farm camps, and that is the first time I met her and learned about WAgN. I had started this list of what we were going to put in our market building. We needed a certified kitchen because we have a daughter who is going to be baking. We needed a place to sell [our produce] because we were sick and tired of going to those farmers' markets. So we started this big, long list of what our building had to be. I went to Haley's place, and she had a studio in her barn. I thought, "Oh, oh! Oh, this could work." Then she had a living room upstairs in her barn like a little office retreat space for her staff. So I started to see that a farm does not have to be just a barn, or a kitchen does not have to be just a kitchen. So it was that that helped me; you know, it encouraged me that we were on the right track to do this. Here she had taken this old barn, and she had an office and a living space and one bedroom and the studio and cows and sheep. It was just very reaffirming that we were on the right track.

A focus group participant noted that the network provided an important opportunity to test the feasibility of ideas with others who have similar operations and interests:

> [I can talk to other women farmers and ask], "Okay, does this really work for you?" and, "Is this what you would do if you only had two hundred dollars to spend? Would you spend it this way?" To really have in-depth discussion with people about what the costs are—that has been the most valuable thing to me and the intimacy of [the network] and field days.

As these examples illustrate, the network functions as an informal educational resource and a space where women farmers feel validated, affirmed, and energized by other women farmers to continue to innovate.

"I'M NOT DOING THIS ALONE": THE NETWORK MITIGATES ISOLATION

Many PA-WAgN members emphasize that being part of the network helps them realize they are not alone. As Joyce notices, "Wow, I am not the only one sitting out in the middle of nowhere, trying and struggling to survive and wanting to be the farmer." May echoes her sentiments: "I have definitely found out there were women in agriculture, and I thought I was the lone sailboat. For me, that was the first big step. It is like, oh my gosh, I am not doing this alone! I know that I felt like I was doing it alone because I don't have a husband, and there are times when it is like, boy, this would be a whole lot easier if there was someone else to bounce ideas off of." Marilyn chimes in, "What I love about WAgN it is the camaraderie, but it is the support and it is being around people who are normal—the women who are normal. You don't meet women like this at home very often." Others said being involved with PA-WAgN has helped them understand some of their earlier struggles. For example, Haley says,

> Hindsight has helped me define some of the tensions and the trouble that I went through during milking cows in the 1980s, and I was a square peg trying to fit into a round hole, and I was a square peg for a number of reasons. I am a square peg primarily because I was a female. I feel like I have interacted with women farmers and, for me, that has just been

> phenomenal. It really has. I am not the only square peg in a round hole. I do feel like I am not the crazy lady that lives down over the hill. . . . I think that [the network] has been very beneficial.

FEELING EMPOWERED

Many women farmers expressed that they feel empowered by the network. As Melody described her experience with PA-WAgN, "I feel empowered here. I feel recognized for who I am and what I am doing when I am with other women farmers, and that gives me strength to go home and keeping what I want to do." Some women explained that they gained confidence in themselves and what they had accomplished on their farms through sharing experiences with others. Sara mentions that she enjoys the "camaraderie that is here":

> I come to these events and there is so much stress at home, and I know the minute I get home it is going to fall back on my shoulders when I walk in the door, but right now, it seems all doable. When I am sitting here, it is like, okay, there is a lot to do, but I am going to do it, and I am going to make it because I am in this environment where everyone has a lot to do, but we are all going to be able to make it. So that sense of just being involved and being able to help others tap into that energy by being a representative here. It means more to me than a lot of other things that I am involved in except for the farm itself.

Sue described herself as previously "feeling incredibly shy and very underpowered. I came to WAgN because I was just learning farming from my partner, and what I found was a bunch of people who were very supportive." About her own farm, she says,

> I was not allowed to say my creative part. . . . You know like that feeling where you were invisible. Yes very invisible, and I think that is what has always kept me coming back to WAgN is that sense of feeling each time I come, I come a little bit more out of myself. I get a little bit more empowered that I am allowed to use my creativity and my ideas. It is safe to say something that might fall flat because it isn't criticized.

FINDING ACCEPTANCE FOR ALTERNATIVE FARMING

Other women mention that they are empowered to make changes in their more conventional modes of farming. "I am working against the conventional farming with the way I was raised on our farm, and so it is nice to have all of the women around me and proof that this is how we can do it. Hey, it is working there. It is working on their farm, and we can use it too." Randy sees the value of the network as a way to understand alternative structures and philosophies of farming:

> I grew up in an area where it was traditional farming, and I worked directly with the very large dairy farmers. I really did not have a concept of the small farmer. I joined PASA [Pennsylvania Association for Sustainable Agriculture], but even then, when I went to the PASA conference, I didn't get the sense of the actual reality of being the actual productive small farm. . . . I moved on and worked full-time and [my partner] more at her regular job and through just her educating me and the connection of other women that this really is a valid career that you can do, and when you look at the benefits not necessarily always as financial.

Marilyn tells the story of teaching a Penn State class where she was not introduced as a real farmer because of her job off-farm, and then, because of that, to make a point, she questioned the students:

> Well how many of one of your parents goes off the farm to work . . . and the realization that farms don't make enough money to support a family anymore. That it is not about being the breadwinners, so to speak. . . . You know we are not out there screaming at our animals like our neighbor is. We are not out there kicking the equipment like our neighbor is. You know it is about connecting with the farm as opposed to trying to force that farm to be something to support us or whatever. That has come from the interaction from WAgN.

Using New Leadership Skills

Women in agricultural networks are also empowered to work with other organizations that serve the interests of women, sustainable agriculture, and their local community. After attending leadership training with PA-WAgN, Liz explains that she became involved in the League of Women Voters because they are a powerful policy and lobbying group. She participated in their caucus on women in agriculture and now serves in a leadership capacity on her local board and goes to state-level meetings. She is trying to make connections between her work with PA-WAgN and the League of Women Voters. She describes women in the league as very powerful women. When she first went to the state convention, she expected "a stuffy group of old women who were working on elections." But rather, she found women who have strong opinions, who discuss their ideas, and she found them very similar to the women in her interactions with PA-WAgN members.

Leslie explores how her involvement on the PA-WAgN steering committee opened up other leadership opportunities for her: "Well, when Ann asked me to be on the steering committee for WAgN, I did not think I had much to bring to it, and at that point I was still rather shy." After participating on the PA-WAgN steering committee, she made other connections and has been asked to be involved in other organizations. "Because of all of the connections I had with the PA-WAGN, I got invited to be on the Buy Fresh, Buy Local steering committee meeting for the Pennsylvania Association for Sustainable Agriculture." She also notes how a person she met through PA-WAgN referred her to the governor's chef, who later called her to get advice about switching to local food.

Dara, a long-standing member of the steering committee, shows how the PA-WAgN leadership training helped her resolve a conflict in her church.

> I was trying to start this huge conflict resolution within our church. There was this huge conflict about putting in a seventy-five-thousand-dollar elevator or not. The church had like sixty members. I was able to take some of that stuff from the leadership workshop and all of those activities that we did there, and it helped to try to resolve that within that church. Yes, it happened within eighteen months. I actually used those skills with helping a whole other group reach some sort of resolution. So I felt like that was very clear and definite, and I learned something here and I used it there.

Leadership skills developed through involvement in the PA-WAgN network have encouraged women farmers to move beyond the local women's network in their organizing and educational efforts.

Conclusion

Cooperative Extension and nongovernmental farming organizations have limited their recognition of women farmers because they do not meet the traditional gender roles for women on farms. Yet these traditional farming organizations don't provide the essential information, technical support, social support, and legitimacy that women farmers need. To fill this gap, women farmers have worked with each other and with individuals in land-grant universities to create women's agricultural networks. As women's agricultural networks serve women farmers, the women in these networks are redefining and expanding the definition of what it means to be a farmer. As part of a transition in agrarian feminism, women's agricultural networks provide a starting place for a structural response to the needs of women farmers.

Chapter 6

From the Ground Up: A Feminist Agrifood Systems Theory

Many women farmers have committed their lives to changing the farming and food system through sustainable agriculture. We derive a new theory of women in agriculture based on our feminist research and praxis (working with women farmers in the Northeast) on women farmers and their reshaping of the food system. These new women in agriculture move beyond agrarian feminism.

In her book *On Behalf of the Family Farm: Iowa Farm Women's Activism Since 1945*, Jenny Devine argues that farm women in Iowa, in their efforts to organize, adopted a strategy of agrarian feminism. She explains that in the first half of the twentieth century, Iowa farm women activists adhered to a politics of social feminism that emphasized political applications of women's roles as caregivers and mothers, solidified by social institutions that provided educational programs in home economics from Cooperative Extension. During this period of social feminism on farms, women emphasized how their traditional caregiving roles could improve life on farms and in the countryside. Devine argues that beginning in 1945, with the decline in farming and rural areas, farm women shifted their rhetoric and action. They "continued to stress their dependence on men, but . . . emphasized working partnerships between husbands and wives, women's work in agricultural production, and women's unique ways of understanding large-scale, conventional farming" (Devine 2013, 3). Devine describes this approach as agrarian feminism because it is based on a politics of dependence in which women recognize and embrace their dependence on men and the family but also use these claims to enter public space and speak on behalf of agriculture and rural communities. These agrarian feminists did not explicitly embrace feminism or necessarily identify as feminists or agrarian feminists, but they used feminist organizing strategies and models of collaboration and cooperation. Although Devine's theoretical perceptions of women on farms in the twentieth century contributes valuable understanding of the women at that time, the theory is based on a twentieth-century paradigm of women on farms as dependent, that is, embodying the role of the farm wife within a heterosexual marriage. Given the current social movement of women entering agriculture who are independent on farms, we put forward a feminist agrifood system theory (FAST) that emerges from the actions of these women in agriculture as they assume, assert, and choose a broader range of roles in agriculture.

Building on Devine's work, we argue that the individual and collective efforts of women farmers are transforming agrarian feminism, providing both a critique and an alternative to the conventional and patriarchal agricultural system. We argue that these women farmers are creating a feminist agrifood system. We draw on our empirical research, our collaborations with these women farmers, and the theoretical literature from sustainable agriculture and feminist theory to craft a theory that is characterized by six interrelated themes: (1) creating gender equality for women on farms; (2) asserting the identity of the farmer; (3) accessing the resources they need to farm, including innovative ways to access land, labor, and capital; (4) shaping new food and farming systems; (5) navigating agricultural organizations and institutions reluctant to meet their needs; and (6) in reaction, forming new networking organizations for women farmers to respond to their needs. These six themes, which we have discussed in detail in the previous chapters, form the basis of a new theory that provides a framework for understanding the shifts in agriculture and women's roles in the US based on data and experience working directly with women farmers in the Northeast. We summarize each of the themes below.

More Gender Equality for Women on Farms

Women farmers are producing crops and establishing livestock enterprises on farms either on their own or interdependently with other women or men, some of whom are their spouses. In fact, many women farmers are transforming gender relationships in farm households and in the broader agricultural community. They are moving toward more egalitarian relationships between men and women in farm households. Some women are married to men who are not involved in farming. Some live in different types of households other than heterosexual partnerships. Some women are single, while others have domestic and farm partners. Many of these women farmers have largely transcended the politics of patriarchal dependence on family farms. They are creating new types of families and new types of farms in which the traditional division of labor and responsibilities between the man as the farmer and the woman as the farm wife is not the only one to exist. At the same time, men who farm with these women, increasingly view them as equal partners.

The gender equality evident on some farms in the northeastern United States differs from, and relates to, the global shift toward the feminization of agriculture. The feminization of agriculture refers to the increasing ratio of women to men in agricultural production and the increasing proportion of women who are employed in agriculture. Women's responsibility for agriculture is increasing, especially in many developing countries, as men migrate or seek other forms of employment. Globally, the best data available suggests that women comprise 43% of the workforce in agriculture. De Schutter (2013) argues that the feminization of agriculture is actually the result of three very different phenomena. In the first case, women take over farming when male adults on farms migrate or take alternative employment leaving women to provide for household food security or the subsistence needs of the family. In the second instance, women take over farming the family land to produce primarily for the market rather than the subsistence needs of their household. The third type of feminization of agriculture involves women's employment as farmworkers, typically in large-scale corporate agriculture enterprises. As De Schutter (2013) notes, each of these forms of the feminization of agriculture results in different sets of gender relationships and different types of agrarian change. Our documented evidence of the move toward gender equity in agriculture in the US, and the Northeast in particular, can be considered yet another form of the feminization of agriculture, but it takes a different form than the agricultural transitions that are occurring in many developing countries. In the transition on smaller scale farms in the US, we see a feminization of agriculture that values women as managers and workers and moves toward more gender equitable relationships in agricultural households and communities. Our results suggest the need for further study on the extent to which these outcomes of emerging feminist agrifood systems exist in other regions of the US or other parts of the world or if women's responsibility for agriculture is undervalued and characterized by exploitation.

Women Assert the Identity of Farmer

Farm women are increasingly identifying as farmers. We argue that identifying as a farmer is a feminist move within the traditional agricultural community and a challenge to longstanding patriarchal relationships in farm households and communities in which women were typically viewed as farm wives. As in Devine's conception of agrarian feminists, few of these women farmers explicitly identify as feminists, but they do proudly identify as farmers. The reticence of women farmers to explicitly identify as feminists is understandable in the complex and contested terrain of agriculture and feminism that they are negotiating. Donna Haraway (1991) insists that although there is no one privileged feminist standpoint, we must look carefully at people's location, position, and embodied experiences that situate their knowledge. There are multiple varieties and perspectives of feminist thought and action, and we are not labeling the women featured in this book or women farmers in general as adopting any one feminist standpoint (Harding 2004). Similar to Devine, we recognize the ways that the women farmers we work with are acting in ways that are consistent with feminist practice, and that is why we describe this theory as feminist.

We argue that women who claim the identity of farmer are making a political move that demonstrates their claim to a form of feminism, whether they explicitly claim the identity of feminist or not. Clearly, women claiming the title of farmer within their households and the broader agricultural community are political moves that implicitly suggests their allegiance with a form of feminism. Rather than merely adhering to a patriarchal model of a farmer, they are creating a new farmer identity. Further research might explore the prevalence of this new farmer identity in other parts of the country and other regions of the world. This research might examine what happens as more women in both sustainable and conventional agriculture claim the identity of farmer. Future researchers might ask, is there contestation about who has legitimacy to speak for women farmers within agricultural communities and organizations?

Women Access Land and Capital

As Devine argues, the politics of dependence for women on farms is steeped in their dependence on men for access to land, capital, and other resources for farming. Traditional patriarchal inheritance patterns have resulted in male dominance in agriculture and farming. As a strategy to resist this longstanding situation, many women farmers have accessed land in alternative ways. Either on their own or jointly with their partners, many women farmers have pursued creative arrangements for accessing land and capital. These alternatives include farming on small land holdings, diversification, farming on public land, and reliance on off-farm income. Women farmers are capitalizing on these alternative sources of land and capital to shape new systems of farming, enabling them to overcome many of the barriers to traditional farming. Here, women are using their creativity and multiple skills to actively seek these alternatives because they want to farm. Potential implications and directions for future research encompass, first, an examination of the extent to which the presence of more women landowners and farmers challenges the broader patrilineal systems. Further questions include, how might new ways of passing on land be created, and what new forms of financing would allow women and others from nonfarm families to successfully enter farming? What types of labor arrangements on farms should feminist agrifood system activists advocate so as not to replicate exploitative and unjust working conditions for farmworkers, farm family members, and apprentices?

Women Shape New Agrifood Systems

Women farmers are shaping new farming systems using strategies that emphasize smaller scale farms, diversified high-value and value-added products and enterprises, unique marketing strategies, and sustainable production practices. Many of these new farming strategies alter the concept of the traditional division of labor between men's productive labor on farms and women's responsibility for reproductive labor. Agriculture is no longer confined to growing crops and raising livestock but includes processing of farm products, collaborative marketing, and on-farm education.

Women's traditional responsibility for reproductive work and food work such as food processing, cooking, and entertaining on the farm has been transformed into valuable labor, management, and a source of profit for farms.

Many women farmers value their relationships with their communities and promote local food systems. They develop relationships with their customers and are committed to providing high quality food. Many encourage people to come to their farms to learn about farming and their production practices. They promote consumer understanding of agriculture and food production with a particular emphasis on local food. In so doing, they are conveying that food is more than a commodity—it is a connection among people. Many women farmers promote local food in the context of broader and international movements such as the food sovereignty movement, slow food, and food justice.

Future research might focus on the potential implications of women's involvement in shaping new food systems. Does the emphasis on women and food perpetuate women's traditional domestic role? What thoughtful balance might be struck to build a food system that values food work? As the farmer Sasha suggested, building community is very difficult for farmers. To what extent are the new relationships they are creating sustainable and positively affecting food systems at a local level?

Women Negotiate Agricultural Organizations and Institutions

Numerous agricultural organizations and institutions exist, but they rarely serve the explicit interests of women as farmers or, in broader terms, promote gender equity in agriculture. In fact, many agricultural organizations have involved farm women as homemakers and farm wives, rarely accepting them as farmers in their own right. Traditional farm organizations such as the American Farm Bureau or large commodity organizations tend to see women farmers primarily as farm wives, as Devine has argued. Some women farmers are involved in smaller commodity organizations or sustainable agricultural organizations, for example, fruit and vegetable growers or sheep and wool growers. However, most of these organizations tend to have very few women in leadership positions, and they do not address the specific concerns of women farmers.

Women Form New Networking Organizations for Women Farmers

As a result of their struggles with existing agricultural organizations, women are taking on leadership roles and forming new types of networks and organizations. These positions define them more broadly than as supporters of their husbands' or male relatives' occupation. Instead, women are coming together to affirm their own role as decision-makers on the farm. In organizations such as the Pennsylvania Women's Agricultural Network, Vermont Women's Agricultural Network, and the Women, Food, and Agricultural Network, women farmers work together to support their experiences as women, farmers, and entrepreneurs. The networks provide programs and activities that facilitate shared knowledge about business and farming practices and provide space for legitimizing their role as farmers. Not all of the members explicitly define themselves as feminists, but they support feminist principles of mutual support, expect equal treatment of women in the agricultural sector, and proudly proclaim their identities as farmers. Future research questions might address the sustainability of women's agricultural networks when long-term institutional support is lacking. How can institutions such as Cooperative Extension better address the needs of women farmers and other farmers outside the traditional patriarchal agriculture system? How will women's leadership in mainstream and sustainable agricultural organizations transform gender relations and other aspects of the agrifood system?

A theory not only explains a phenomenon, but helps to predict an outcome. What then can this new theory, FAST, suggest for the future? To summarize on an abstract level, we have studied one group that has attempted to break into the structure of agriculture as it existed. The group is different in a major demographic characteristic than the group in the past: gender. And this new group wants to farm in different ways than those established already in the structure. Further, the group trying to break into the structure and their type of farming are not recognized as necessarily acceptable or legitimate, so much so that agricultural institutions and agencies designed to help all farmers have resisted helping this new group with what they need to succeed, such as education and funds. The identities of members of the group trying to break into this structure are, on the one hand, strong, in that they think of themselves as legitimate farmers, but that view is not shared by others in the overall agricultural structure. Given this outside group of farmers is of a different demographic, here gender, than the group who dominates the agricultural structure, what does the feminist agrifood systems theory based on empirical data suggest for such a group to succeed?

For members of a group to enter and transform an existing agricultural structure successfully, they must possess the qualities that we found: a strong determination to farm; use of innovative measures to secure resources that include land, labor, and capital; and awareness and action by academic leaders of the same demographic (here, women) at a university, perhaps, to recognize the isolation and challenges facing the new group of farmers and to bring them together to talk about their need to help each other. What also helps a new group penetrate an older agricultural structure is response by members of the new group to a need assessment process, educational programs, and a call to network and mentor each other. With these values and social processes in operation can come (1) the success that flows from education, networking, and leadership; (2) stronger identity and legitimacy as a farmer; and (3) a new kind of agriculture established side by side with the old so that agriculture, overall, becomes redefined.

Should another group with a different demographic struggle to enter and transform the mainstream agricultural structure, we can draw on FAST to begin to predict their ability to sustain themselves. We can also observe other processes that other new groups may develop to enter and transform the mainstream structure of agriculture.

FAST in Context

This study of women farmers and the development of FAST provides a useful point of comparison for other regions and populations to add to our knowledge of agricultural development. As the feminization of agriculture has progressed in the global context, a critical question is whether feminist agrifood systems are thriving as alternatives to conventional agriculture in other regions of the US and the world. Many women farmers are pushing not only for their place as farmers but for broader social justice, access to healthier food, and environmental justice in their communities. We recognize that we have studied women farmers who pursue a particular type of agriculture in the northeastern United States and that they may differ substantially from women engaged in other types of agriculture. Some examples might include large-scale commodity farms, farms far from urban centers and with little access to direct markets, European farms, or even smaller scale farms in developing countries. To measure the broader reach of feminist agrifood systems, researchers might assess differences for women among these types of farms and to what extent they are transforming the current structure of agriculture by claiming the identity of farmer, by changing gender roles on farms, by influencing agricultural decisions, and by networking with other women.

Contemporary feminist theory and activism focus on intersectional issues such as gender, race, class, ethnicity, and sexuality. These issues are at the core of social justice in agriculture. With a few exceptions, the vast majority of women with whom we have interacted in the women farmers' networks described in this book are white. This is not surprising given the small numbers of male and female African American and Hispanic commercial farmers in the Northeast and the Midwest. For example, the overwhelming majority of farmers in Pennsylvania are white, while only 0.1% (103) are African American and 1.0% (652) are Hispanic. Some urban women farmers in the PA-WAgN are African American and grow for, or have educational programs directed at, African American audiences. PA-WAgN helped establish a farmers' market in a designated food desert and provided a point-of-sale machine to accept SNAP cards from low-income customers, and it organized cooking classes with local community members to teach people how to use healthy and fresh foods. But much more effort is needed to engage farmers and eaters from diverse socioeconomic, racial, and ethnic backgrounds. With this in mind, we acknowledge that FAST is not a conclusion but rather a tool we offer that we hope will be used to better understand women in agriculture more thoroughly. Despite these limitations, the women farmers that we discuss in this book have moved beyond agrarian feminism. They are crafting a model of agricultural production and food provision, which we describe as feminist agrifood systems, that strives to be environmentally sound, socially just, and transformative for rural gender relations. In these ways, they are standing on new ground and growing sustainable agriculture.

APPENDIX Methodology and Data Collection: Linking Research to Pedagogy, Action, and Policy

In the ten years that we have worked with PA-WAgN, we have been committed to bridging research, pedagogy, and action for women farmers with the long-term goal of social transformation. This approach is part of a larger tradition of feminist action research that emphasizes participation in research with the goals of challenging patriarchy, changing power structures, and empowering individuals and communities in a range of areas such as development, land rights, and violence against women (Abraham and Purkayastha 2012; Reid 2004). Our overarching intention is to be meaningful and transformative to the lives of women farmers in Pennsylvania and beyond. This is the reason we have employed the methods used in this work.

In this appendix, we describe our research methods in greater detail than is possible in the book itself. The methodologies we used to collect data reflect our goals of improving the conditions for women farmers and empowering them by providing opportunities for women farmers to tell their own stories about their experiences. We have used a range of methodologies to collect data from multiple groups, including women farmers and those who work with women farmers. The table provides a summary of these approaches, and the text below describes these approaches in greater detail. As noted in the preface, we identify each data collection with a brief reference code derived from its title and use it in the text as a shorthand.

Focus Groups of Women Farmers (FGWF)

As we were beginning our research and developing PA-WAgN's educational programming, we conducted focus groups to determine the educational needs of women sustainable farmers in Pennsylvania. We chose focus groups because they allow the respondents to interact with and build off the

Table A.1

Title	Data Collection Method	Date(s) of Data Collection	Research Participants	Number of Participants	Location
Focus Groups with Women Farmers (FGWF)	Focus groups	2005	Women farmers in Pennsylvania	28	5 locations in PA
In-Depth Interviews with Women Farmers (IIWF)	Interviews	2005–2006	Women farmers in Pennsylvania	22	Women's farms
Educational Needs Assessment with Women Farmers (ENAWF)	Surveys	2006–2007	PA-WAgN members	151	NA
Evaluation of Educational Events (EEE)	Surveys	2006–2009	PA-WAgN event attendees	452	Multiple
Survey of Northeast Women Farmers (SNEWF)	Surveys	2007–2008	Women farmers in the Northeast Region	815	NA
Focus Groups with PA-WAgN Steering Committee (FGSC)	Focus groups	2012	PA-WAgN steering committee members	15	Conference center
Survey of Cooperative Extension Educators (SCEE)	Surveys	2007	Penn State Cooperative Extension educators	115	Online
Interviews with Cooperative Extension Educators and Specialists (ICEE)	Interviews	2007	Penn State Cooperative Extension educators	18	Campus and telephone
Research Team Observations (RTO)	Participant observation	2004–current	Women farmers, steering committee members, event attendees	NA	Multiple event sites

ideas of one another, generating richer and more detailed data than a single interview with one person (Bloor et al. 2001). Focus groups have also been used as a way to empower women through networking (Pini 2002). The research method brought together women who did not know each other but lived in the same region for an evening of discussion about their farm operations and networking, and in addition, they reported to us their preferences for educational programs. We were quite surprised when the participants at all focus groups expressed their delight at getting together and asked for more focus group discussions to explore issues they faced, thus echoing Pini.

We conducted five focus groups, one in each of the five regions in Pennsylvania in 2005. Individuals on the research team conducted all the interviews. The discussion focused on identifying the educational needs of women farmers, including topics, pedagogical methods, and the context in which women preferred to learn. We transcribed the focus group interviews, and two of the authors coded the transcripts separately, using codes including values, motivations, educational content, experiences with educational contexts, organizations, sources of information, barriers to seeking and obtaining education, and networking (as observed during the focus group). In addition, each focus group participant completed a participant profile describing their identities as farmers, the type of farm operation, and other characteristics of their farm.

Twenty-eight women participated in the focus groups, which ranged in size from four to nine participants. The participants, all white women, ranged in age from thirty-five to fifty-four years. Nine women identified themselves as the sole operator of their farms, eight as one of the main operators, eight as a farming partner, two as farm helpers, and one did not respond. The average number of years in farming was ten, with a range between one and thirty-three years. The median number of years in farming was five. Eleven of the women identified their farming operation as sustainable; seven as certified organic; seven as organic but not certified; one as conventional; one a mix of conventional, sustainable, and organic; and one did not specify. They operated a wide variety of enterprises with fifteen fruit and vegetable farms; eight mixed livestock, fruits, and vegetables; three livestock only; two dairy; and two "other" enterprises whose products included maple syrup and sheep wool. They used diverse marketing methods with twenty-one employing two or more types of marketing, including direct, retail, wholesale, preorder, and subscriptions. Twenty-four used some form of direct marketing.

In-Depth Interviews with Women Farmers (IIWF)

We conducted interviews with women farmers in Pennsylvania to understand what type of educational programs they were interested in, what barriers they faced as women farmers, and what they considered their successful entrepreneurial activities. We used in-depth interviews as a way to gather rich data on the lived experience of women farmers (Patton 2002). In-depth interviewing is an excellent strategy for "getting at the subjugated knowledge of the diversity of women's realities that often lie hidden and unarticulated" (Hesse-Biber 2007, 111).

The primary selection criteria for participation in the interviews included the operation of a successful small- to medium-sized farm (farm sales less than $250,000) and some recognition for this success among educators, organizational personnel, and peers who recommended them for the interviews. Because the research was supported by a grant from the USDA to help the agency support small- and medium-sized farms, we sought a purposeful sample of successful entrepreneurs with a variety of operations, including small grains, dairy, fruit and vegetables, and mixed operations. We conducted in-depth, semi-structured interviews with twenty-two women farmers in the winter of 2005–2006. We interviewed ten women who farmed independently and twelve women who farmed in partnership with their spouses or others. Most (fifteen) of the women were members of PA-WAgN. All of the women who we contacted agreed to participate in the study. The interviews were conducted at the respondents' farms and were one and a half to two hours in duration.

Two individuals on the research team conducted all the interviews and met regularly to review the conduct of the interviews. We asked respondents to tell us about their entrepreneurial activities, their marketing and livelihood strategies, their use of sustainable agricultural practices, the type of information and educational programs they sought, barriers to successful farming, and their involvement with agricultural organizations and networks. We transcribed the interviews using codes for farming history, production practices, marketing strategies, education, organization involvement, barriers, opportunities, gender identity, livelihood strategies, current and future needs, sustainability, decision-making, motivations, and community. The coding consisted of an iterative process designed to identify common and recurring themes as well as outlying trends. Two of the authors coded the transcripts separately to enhance the reliability of the analysis.

Educational Needs Assessment with Women Farmers (ENAWF)

To understand the degree to which the needs identified in the focus groups extended to other women farmers in Pennsylvania and whether other needs existed that did not surface in the focus groups, we conducted a systematic survey to determine the educational program needs of women farmers in Pennsylvania in 2006 and 2007. We considered an educational need as a discrepancy between an audience's current status and some desired result (Wilkin and Altschuld 1995), the working assumption being that education can bring about the desired result. The survey instrument was comprised of closed- and open-ended questions that assessed skill levels; perceived barriers to farm success; need for, access to, and utility of educational programs; preferences of content and format of educational programs; and demographics of the respondents.

We distributed the survey to all women farmers who registered as members of PA-WAgN (online or by mail) or who attended a PA-WAgN educational event. Approximately 700 survey instruments were distributed between February 2006 and September 2007; of these, 151 needs assessment surveys were returned (Barbercheck et al. 2009).

Evaluation of Educational Events (EEE)

From 2006 through 2009, we systematically evaluated educational events sponsored by PA-WAgN using an end-of-event paper evaluation survey. Although we had conducted fifty-five events, we focused the evaluation on the thirty-seven that were solely sponsored by PA-WAgN. From the evaluations, we obtained data on demographics to see if we were reaching our target audience, on educational impact to ascertain if we were achieving our educational goals in terms of knowledge and intention to make changes within two years, on the expansion and enhancement of the network to ascertain the nature and degree to which the educational events had an impact on the network, and on marketing issues to see how Cooperative Extension might improve women's participation at Extension events. To measure the demographics, we asked about age, occupation, location of residence or farm, and gender, the latter to separate the data of a small but increasing number of men who came to the events. To measure knowledge and intention change, we used retrospective before and after questions with four answer categories, these retrospective questions having been used in the evaluation of many Pennsylvania extension programs (Swackhamer and Kiernan 2005). To measure the impact on the network and marketing, we drew on concepts that had emanated from the focus groups, including using the network for business contacts, social support, and innovation. Of the 891 people attending these PA-WAgN events, 452 participated in the four years of evaluation (51%) (Kiernan et al. 2012).

Survey of Northeastern US Women Farmers (SNEWF)

To understand the extent to which our interviews, focus groups, and survey need assessment findings represented the experiences of a broader population of women farmers, we conducted a survey of women farmers in nine northeastern states. The survey was administered in the winter of 2007–2008 in Pennsylvania, New York, Maine, New Hampshire, Connecticut, Massachusetts, Vermont, New Jersey, and Rhode Island. For this study, we defined women farmers as women who live on a farm and participate in farm labor or decision-making. Because no publicly available list of women farmers existed, we developed a sampling frame by drawing from multiple sources, including membership directories, advertising materials and brochures, and websites of farming organizations. Organizational sources represented multiple agricultural sectors, including traditional commodity producers (e.g., state-based labeling programs), sustainable or organic producers (e.g., Northeast Organic Farming Association, Pennsylvania Association for Sustainable Agriculture), and women's agricultural organizations (e.g., Vermont WAgN and Pennsylvania WAgN). Farms included in the sampling frame produced products prevalent in the region, including dairy, grains and oilseeds, horticulture, livestock and livestock products, fruit, greenhouse and nursery, specialty and value-added products, or other farm-based enterprises (e.g., bed and breakfasts, corn mazes, tours, educational programs).

Surveys were mailed to a sample of 2000 farms drawn randomly from this sampling frame. A modified Dillman method (2000) was used to administer the mail survey, including a five-contact mailing procedure (prenotification, survey, reminder postcard, reminder with a survey, and a final reminder letter) and a one-dollar incentive payment. Labels and letters were addressed to the female name in the sample. If a female name was not available, the label was addressed to the "Female Household Member" at the farm address. A total of 815 completed surveys were received, for a response rate of 40.7%.

Survey questions included information about the characteristics of their farm, types of production strategies, marketing strategies, farm identity, labor participation on the farm, farm decision-making, and demographic characteristics. Survey participants represented all northeastern states, with about two-thirds from Pennsylvania, Maine, Vermont, New York, and Massachusetts. They were primarily Caucasian (89.0%) and the average age was fifty-one years old. The farms operated by the respondents were small, with almost two-thirds (63.3%) reporting gross farm sales below $50,000. Most farms were small in scale, with a median total acreage of 70 acres (owned and rented) and a median of 14.0 animal equivalent units. Nearly three-quarters (73.9%) produced row or horticultural crops, while two-thirds (66.4%) produced livestock. The primary crops grown include vegetables (52.6%); small fruits and brambles (30.7%); pasture (27.6%); horticulture plants, nursery, or flowers (25.0%); and alfalfa or hay (24.4%). The primary livestock on the farms are laying hens (36.7%), horses (26.7%), beef cattle (21.6%), and hogs (21.2%). Respondents primarily sold directly to consumers, with 71.7% using direct retail outlets or direct sales to consumers. Similarly, 74.6% used at least one strategy to add value to their farm products, such as organic and specialty production, on-farm processing, or agritourism. The women were highly educated, with more than half (56.9%) with college degrees or higher. Less than half of the women (43.3%) lived in households with a total household income less than $50,000. However, as noted above, almost two-thirds operated farms with less than $50,000 gross income. Therefore, it is not surprising that the majority of the households (73.9%) had income from nonfarm wage and salary jobs.

Focus Groups with PA-WAgN Steering Committee Members (FGSC)

We conducted four focus groups with fifteen PA-WAgN steering committee members during the 2012 annual retreat. Two of the four focus groups concentrated on their experiences as members of PA-WAgN, as members of the steering committee, and as participants in the leadership training conducted during the retreat. These focus groups also explored how participating in the steering committee and PA-WAgN has affected them in terms of being a farmer and a woman. The other two focus groups explored how steering committee members experienced mentoring others or being mentored as farmers. Two of the book authors conducted the focus groups. The sessions were audio-recorded, transcribed verbatim, and coded for themes related to the questions listed above, particularly related to PA-WAgN impact described in chapter 6, including legitimacy, mitigating isolation, entrepreneurship, and leadership. These focus group conversations were especially helpful for developing the mentoring program later implemented by PA-WAgN.

Survey of Cooperative Extension Educators (SCEE)

Women farmers consistently reported frustration with accessing and working with Penn State Cooperative Extension educational resources and some personnel. Our experiences also indicated a disconnect between the women farmers with whom we were working and the types of and methods used in the educational programs offered by Cooperative Extension. To characterize this gap and identify the reasons for its existence, we conducted research to understand how Cooperative Extension educators and specialists perceive their work and interaction with women farmers and how they describe women farmers and their educational needs.

We developed and administered an online survey (via SurveyMonkey) to all county-based educators in within PSU Cooperative Extension during the winter and spring of 2007. A total of 260 personnel from all program areas were notified of the survey by extension administrators in the college and received four reminders through email by state specialists (Dillman 2000). A total of 115 educators responded, for a response rate of 44.2%. The survey provided the following context and rationale:

> The USDA's definition of a farm is "agricultural places that produce and sell, or would normally sell, $1,000 or more of agricultural products." USDA research indicates that the number of women farmers is increasing

> each year. The Pennsylvania Women's Agricultural Network (PA-WAgN) has developed a survey to obtain your opinion about, and experiences with, women farmers in your Extension region. Your views are important for focusing programs and materials for women farmers.

In the survey, educators were asked a series of questions to assess their level of knowledge about women farmers in their region and their perceptions of women farmers' responsibilities, educational needs, and challenges. Educators were also asked about their program delivery methods, marketing practices, and demographic information (e.g., region, program area, and gender). Questions on the survey were based on concepts drawn from a review of previous research on women farmers (Danes 1996; Swackhamer and Kiernan 2005; Trauger et al. 2008; Willits and Jolly 2002). Most of the questions were closed-ended; an initial open-ended question asked educators to describe women farmers' educational needs so as to not bias what was most relevant at that time. Six Extension educators (male and female), representing the target audience for the survey, reviewed an initial draft, evaluating the survey for comprehensiveness, acceptability of language, and relevancy.

Interviews with Cooperative Extension Educators and Specialists (ICEE)

To further understand the issues about the perception of women farmers raised in the surveys, we conducted semistructured interviews with a subset of Extension specialists and educators. The interviews focused on their attitudes, beliefs, and opinions about any women farmers with whom the educators and specialists had worked. The individuals selected were programmatic leaders who had direct interaction with, and influence over, the types of programming developed and delivered to audiences across the state. We interviewed eighteen men and two women, including county-based educators and campus-based extension specialists. The interviews were conducted by two of the authors over the phone and in person during the summer of 2007. The interviews were audio-recorded, and the interviewers also took detailed notes and captured key phrases and quotations during and immediately following the interview as back up, should the tape falter. Transcripts and notes from each interview were shared among the six research team members who read all the interviews and generated lists of themes cutting across all the interviews. The final list of codes included the following concepts: perception of the definition of farmer, definition of farm, ideas of women on farms, educational programs, and definitions of equity and affirmative action. Each interview was then coded by two authors who discussed any differences in coding and collectively identified the final code in order to increase reliability. Members of the research team then created and shared summary documents for each code that highlighted relevant subthemes and related narratives from the interviews.

Research Team Observations (RTO)

From 2004 to 2014, researchers involved in this project have worked with women farmers in various capacities. Throughout this time period we have used the findings from our research to design our educational and networking activities and to build the capacity of PA-WAgN as an organization that better serves the interests of women farmers. We have attended and organized more than 118 on-farm events, field days, webinars, and conferences with a total of 3,542 participants. We have talked to the people who attended events and participated in the dialogues; we used this information to improve our educational programming. We have attended and organized eighteen in-person steering committee meetings and fifteen telephonic steering committee meetings. Annually, from 2004 to 2015, we have convened a two-day steering committee meeting where members share information about changes in their farm operations, in their personal lives, and in their political involvement during the previous year. At these two-day events, we have come to know many of these women farmers very well. These ongoing formal and informal interactions with farmers, especially with PA-WAgN steering committee members, have enhanced our understanding of the joys, frustrations, challenges, and personal experiences of these women farmers. Through participation in events and activities of the network, we have come to understand more deeply the ongoing concerns and possibilities for women farmers.

Feminist Practice within the Research Team

This book is the result of a collaborative effort. We experience our collaboration as a feminist strategy because as Pratt argues, it is both a way of situating knowledge and a source of intellectual and social support (Pratt 2010). Over the course of ten years, our research team, consisting of faculty members, postdoctoral researchers, and graduate students, met regularly to plan research, analyze results, and write papers, articles, and this book. We work across disciplinary boundaries. Three faculty members are rural sociologists (one is in rural sociology and women's studies), one is a program evaluator, and one is an entomologist. We have collaborated with a postdoctoral researcher who is a geographer and a number of graduate students who have all been rural sociologists, with some also obtaining dual-title degrees in women's studies. Some have strengths in qualitative methods, while others have strengths in quantitative methods. This blend of methodological strengths expands the methods available to us and strengthened our work. We endeavor to conduct our work as a feminist collaborative team within a college of agriculture at a land-grant university. Initially, we plan our research together, discuss how to analyze our data, look for key findings, and debate what our findings mean. Subsequently, we share manuscripts to build on each other's ideas, and together reflect on what our research means for educational programming for women farmers and how our research contributes to broader scholarly research. We acknowledge the power differentials between professors and graduate students, and we do have divisions of labor, but we are committed to listening to each other and valuing the scholarly contributions of each participant. Our method of writing this book has also been collaborative, in which the group has worked together to hone and shape the basic argument and structure of the chapters. Each author has taken turns commenting on and revising each of the chapters. Finally, we support each other as we make our way through our daily work as faculty members and graduate students.

FUNDING SOURCES

"Building Sustainability for New and Beginning Women Farmers through Peer Learning, Mentoring, and Networking." USDA New and Beginning Farmer and Rancher Development Program. 2009–2015. Grant # 2012-49400-19602.

"Risk Management for Northeast Women Farmers: Adapting Annie's Project." USDA Northeast Center for Risk Management Education. 2007–2009.

"Sustaining Small Farms and Rural Communities: The Role of Women Farmers." National Research Initiative of the Cooperative State Research, Education and Extension Service, USDA, Small Farms and Rural Agricultural Communities. 2005–2008. Grant # 2005-55618-15910.

"WAgN: Sustainable Agriculture Network by and for Women Producers." USDA Northeast Sustainable Agriculture Research and Education. 2005–2008. Grant # LNE05-226.

NOTES

CHAPTER 1

1. The US Department of Agriculture (USDA) National Agricultural Statistics Service (NASS) conducts the Census of Agriculture on a five-year cycle in years ending in 2 and 7. Although the Census of Agriculture has been conducted since the early 1800s, only since 2002 did the Census allow reporting of more than one farm operator per farm. http://www.agcensus.usda.gov/Publications/2012/Online_Resources/Highlights/Farm_Demographics/.

2. According to NASS, in 2012, of the 2.1 million principal operators in the United States, 288,264 were women, and represented a 6% decrease since 2007. But for all female operators (principal, second, and third), the decrease was only 1.6%. http://www.agcensus.usda.gov/Publications/2012/Online_Resources/Highlights/Farm_Demographics/#fewer_women.

3. Sustainable and organic agriculture are growing sectors in US agriculture, particularly among women farmers. Although often thought of as interchangeable, they have distinct meanings. The term "sustainable" encompasses agriculture broadly, and considers environmental, economic, and social outcomes. In the US, organic products are labeled through an accreditation and audit process defined by the USDA that indicates that food and other agricultural products have been produced through methods that integrate cultural, biological, and mechanical practices that foster cycling of resources, promote ecological balance, and conserve biodiversity. The national standards on organic agricultural production and handling specify practices and substances that may be used in production, processing, and handling of agricultural products sold as organic. Although these definitions differentiate sustainable and organic agriculture, both presume particular philosophical commitments about agriculture and the environment (Gold 2007). At their roots, both were informed by a desire to reduce the negative impacts of pesticides and other inputs and practices common in commercial agriculture, and support building biodiversity and soil health to improve productivity and produce healthy plants and animals, including humans. However, there is currently no government-based certification program for

sustainable agriculture, and thus the use of "sustainable" practices is interpreted very broadly, and often in conflicting ways, by different agricultural constituencies.

4. The percentage of farm household income coming from farming depends on farm typology. Farm income contributes little to the annual income of farm households operating residence farms, is a secondary source of income for households with intermediate farms, and is a primary source of income for those operating commercial farms. The median farm operator household consistently incurs a net loss from farming activities, which means that most farm operator households rely on off-farm income to sustain them. Of the total off-farm income earned by all farm operator households, the majority comes from wages and salaries, followed by transfers (e.g., Social Security) and nonfarm businesses. http://www.ers.usda.gov/topics/farm-economy/farm-household-well-being/farm-household-income-%28historical%29.aspx#.UolC7MeT5CM.

5. The USDA Economic Research Service (ERS) divides US farms into relatively homogeneous groups, or typologies, based on gross farm sales, and defines "family farms" as any farm for which the majority of the farm business is owned by individuals related by blood, marriage, or adoption, or farms organized as proprietorships, partnerships, and family corporations that are not operated by a hired manager. By the ERS definition, 88% of US farms are "small." Small farms are those that have gross annual sales of less than $250,000, while commercial farms have gross sales greater than $250,000. In 2012, 75% of all farms (family and non-family) had gross annual sales of less than $50,000; 13% had gross annual sales of $50,000–$249,000; 8% had gross annual sales of $250,000–$999,000; and only 4% had gross annual sales greater than $1 million (but accounted for 66% of all sales). In 2012, 91% of women's farms had gross annual sales of less than $50,000; compared to 75% for all farms. http://www.agcensus.usda.gov/Publications/2012/Preliminary_Report/Highlights.pdf (USDA 2014c). Farm size is also described by acreage. In 2011, mean farm size for both men and women was 234 acres. 83% of cropland was on farms that were larger than the mean size, and 71% was on farms that were more than twice the mean. The midpoint acreage—where half of cropland is on larger farms and half on smaller—was 1,100 acres. www.ers.usda.gov/publications/err-economic-research-report/err152.aspx.

6. See USDA (2014b), http://www.agcensus.usda.gov/Newsroom/2014/02_20_2014.php. Provides a summary of the changes in the number of farms by size category between the 2007 Census of Agriculture and the 2012 Census of Agriculture.

7. Between 2007 and 2012, agricultural production costs increased 36%. The largest expense categories in 2012 were feed, livestock and poultry purchases, fertilizer, hired labor, and cash rent. The largest percentage increases were in seeds, chemicals, and cash rent. www.agcensus.usda.gov/Publications/2012/Online_Resources/Highlights/Farm_Economics/#snapshot_sales.

8. The 2012 Census of Agriculture showed an 8.2% decline in the total number of dairy farms from 69,890 in 2007 to 64,098 in 2012. However, there was only a 0.1% decline in the total number of milk cows from 9.266 million to 9.252 million during the same period (USDA 2012e). In 2013, the estimated total number of milk cows in the US was 9.221 million with an average production of 21,811 pounds of milk. The average milk cow herd increased in size to 144.3 head in 2012 from 132.5 head in 2007 (Erickson 2014).

9. Three types of producers operate hog and pig farms: independent growers raising hogs and pigs for themselves, contract growers raising hogs and pigs for someone else, and contractors using contract growers to raise some or all of their hogs and pigs. In 2012, independent growers operated 85% of hog operations, but accounted for only 46% (91.4 million) of the 199.1 million hogs and pigs sold. In contrast, corporations were 8% of hog farms and accounted for 34% of sales. Contract growers accounted for 44% (88.1 million), and contractors for 10% (19.6 million) of sales (USDA 2014d).

10. In 2007, about 45% of women operators specialize in grazing livestock, i.e., beef cattle not in feedlots, horses, and to a lesser extent sheep and goats. These specializations account for only 16% of sales by women-operated farms. Farms specializing in poultry, specialty crops, grains and oilseeds, or dairy account for the bulk of sales by women-operated farms (72%), but make up only a 21% share of all women-operated farms. Although the largest share of women farmers specialized in beef cattle farming, only 11% of the US beef farms were operated by women in 2007. Women operated a particularly high share of horse farms, increasing from 17% in 1982 to 31% in 2007. Between 1982 and 2007, the number of horse farms in the US tripled, but those operated by women increased sixfold (USDA 2009b).

11. Another factor contributing to the much smaller percentage of women operating larger scale commodity enterprises may be that women on larger scale conventional farms are less likely than women on smaller operations to self-report as farm operators on census forms. They may be less likely to identify as farmers and more likely to view themselves as partners, helpers, farm wives, or off-farm workers.

12. USDA (2012e), 2012 Census of Agriculture, http://www.agcensus.usda.gov/Publications/2012/Preliminary_Report/Highlights.pdf. Table 56 shows selected farm characteristics of women principal farm operators in 2007 and 2012 including farm size, tenure status, market value, and type of commodity produced.

13. USDA (2015a), "Extension," http://nifa.usda.gov/Extension/. This website describes how Cooperative Extension has changed overtime.

14. USDA (2015c), "US Statistics on Women and Minorities on Farms and in Rural Areas," http://www.nal.usda.gov/afsic/pubs/agriwomen.shtml.

15. Agricultural producers have a number of support programs available to them.

Programs may be tied to specific commodities or to whole farm revenues; they may provide support for current losses or be based on historical production, or they may be in the form of multiyear contracts for conservation. For example, direct payments are available to producers with eligible historical production of wheat, corn, grain sorghum, barley, oats, upland cotton, rice, soybeans, other oilseeds, and peanuts. Producers enroll annually in the program to receive payments based on historical program yields and acreage. Fixed direct payments are not tied to current production or prices and do not require any commodity production on the land, (USDA 2015d).

16. USDA (2015e). USDA Settlements and Claims Processes website includes information regarding claims processes and settlements for African American, Hispanic, Native American, and women farmers and ranchers. http://www.outreach.usda.gov/settlements.htm. Accessed September 23, 2015.

17. For a detailed data on the increase in organic production, see these websites: USDA (2012f), "USDA Releases Results of the 2011 Certified Organic Production Survey," http://www.nass.usda.gov/Newsroom/2012/10_04_2012b.asp; and USDA (2012g), National Agricultural Statistics Service report on "Organic Production," http://usda.mannlib.cornell.edu/MannUsda/viewDocumentInfo.do?documentID=1859.

18. Nevertheless, some contestation exists over exactly what constitutes local food, whether local food is necessarily produced in a more sustainable fashion, and who benefits (Hinrichs 2003; Hinrichs and Allen 2008). The boundaries of "local" vary depending on circumstances in communities, counties, states, or broader geographical regions and can extend from the surrounding community to hundreds of miles. Also, questions remain about the possibilities that local food systems can reach their goal of achieving environmental and social justice aims, including gender equity, while working in the broader context of the agro-industrial food system.

19. A CSA consists of a community of individuals who pledge support to a farm operation with the growers and consumers providing mutual support and sharing the risks and benefits of food production. Typically, members or "share-holders" of the farm pledge in advance to cover the anticipated costs of the farm operation and farmer's salary. In return, they receive shares in the farm's bounty throughout the growing season. Members also share in the risks of farming such as droughts or crop failures. By direct sales to community members, who have provided the farmer with working capital in advance, growers receive better prices for their crops, gain some financial security, and are relieved of much of the burden of marketing (DeMuth 1993). Further, the farmers are in constant contact with the consumers.

20. US Bureau of Labor Statistics (2014), Women in the Labor Force: A Data Book, Report 1049, http://www.bls.gov/cps/wlf-databook-2013.pdf.

21. White House Council on Women and Girls, http://www.whitehouse.gov/administration/eop/cwg/data-on-women.

22. For example, Slow Food, http://www.slowfoodusa.org/.

CHAPTER 2

1. See appendix.

2. See appendix for details of research methods used to collect data presented in this book. Acronyms are used for easy reference. In this case SNEWF refers to the Survey of Northeastern US Women Farmers and IIWF refers to In-Depth Interviews with Women Farmers.

3. In a survey of farmers in a five-state region of the Midwest, 23% of the respondents involved in tractor-related injury events were women (Carlson et al. 2005). Males have been identified as being at greater risk of agricultural injury in general (McCurdy and Carroll 2000) and specifically for tractor-related injuries (Lee et al. 1996). Carlson et al. (2005, 262) suggest that males may be at greater risk of tractor-related injury than females due to greater exposure time or involvement with more hazardous tasks. In these studies, the authors did not segregate results based on whether women were farming alone as a principal operator or together with a partner.

4. See a photo and description of an innovative tractor hitch at Green Heron Tools (2015), http://www.greenherontools.com/products_tractor_hitch.php.

5. See National Public Radio (2012), "Women, Hispanic Farmers Say Discrimination Continues in Settlement," http://www.npr.org/templates/story/story.php?storyId=164833428.

CHAPTER 3

1. Value-added products are defined as products that, as a result of changes in physical state or the manner in which the agricultural commodity or product is produced and segregated, the customer base for the commodity or product is expanded and a greater portion of revenue derived from the marketing, processing, or physical segregation is made available to the producer of the commodity or product.

CHAPTER 4

1. High tunnels are unheated, plastic-covered hoop greenhouses used to extend the vegetable production season in temperate climates.

2. Mob grazing is a style of management that involves moving cattle between small paddocks sized to match the number of grazing cattle and split by portable electric fencing. The goal is for every plant in the grazing paddock to be either eaten or walked on and trampled. Grass in each paddock then rests for 60 to 120 days or

more. This style of management can have positive impacts on pasture productivity and soil quality.

3. Sustainable and organic agriculture are often discussed in opposition to a "conventional" agriculture that emphasizes maximized productivity, economies of scale, and short-term profitability through the use of industrial technologies and specialization, and that deemphasizes long-term environmental and social impacts as "externalities" (Pretty 2008).

4. Based on estimates from twenty-five monitored states, herbicides were applied to 98% of acres planted to corn (USDA 2011a). Fungicides and insecticides were applied to 8% and 12% of acres planted to corn, respectively. Farmers applied herbicides to 98% of acres planted to soybean, and insecticides and fungicides to 18% and 11% of soybean acres, respectively (USDA 2013c). In 2013, 85% of all corn and 93% of all soybeans grown in the US were genetically modified to be herbicide-tolerant, and 76% of all corn grown in the US in 2013 was genetically modified to produce insecticidal proteins. According to one recent estimate (Benbrook 2012), herbicide-tolerant crop technology has led to a 527 million pound increase in herbicide use in the US between 1996 and 2011, while insecticidal crops have reduced insecticide applications by 123 million pounds.

5. According to Gillespie et al. (2010), "Percentages of US farms adopting rBST were estimated at 9.4% in 1996 (US Department of Agriculture [USDA] Animal and Plant Health Inspection Service [APHIS], 2003), 17.3% in 2000, 15.2% in 2003 (USDA APHIS 2003), 16.6% in 2005, and 15.2% in 2007 (USDA APHIS 2007). Larger farms have (had) greater use, esp. farms with more than five hundred cows. In 2007, large farm use rate about 43%."

CHAPTER 5

1. Farmers from different commodity groups (and other representatives of the local community) comprise the local county Cooperative Extension board of directors or advisory board, which sanctions the educational programs that the local Extension agents deliver each year. Budgets, either all or a portion, to support these programs may come from local governments and may be influenced by the local Extension board of directors or advisory board.

2. USDA (2015a), "Extension," http://nifa.usda.gov/Extension/. This website describes current efforts of Cooperative Extension.

3. "Annie's Project Mission Statement," Annie's Project, Iowa State University, http://www.Extension.iastate.edu/annie/mission.html. Also see http://anniesproject.org/.

4. See http://www.rma.usda.gov/news/2013/10/2013outreach.pdf for Annie's Projects funded by the USDA Risk Management.

5. USDA (2015f), "Secretary Vilsack's Efforts to Address Discrimination at USDA," USDA Office of the Assistant Secretary for Civil Rights, http://www.ascr.usda.gov/cr_at_usda.html.

6. But as the USDA civil rights report suggests, limited funding cannot be an excuse for inadequate targeting of programs and funding to women, minority, and limited resource farmers (USDA 1997).

7. The Women's Leadership Commmittee of the American Farm Bureau has held several conferences, and their activities and programs can be seen on their brochure "Growing Strong: American Farm Bureau Women's Leadership Program" at American Farm Bureau (2015), http://www.fb.org/assets/files/programs/women/FB13_92_WLC_Brochure_f_web.pdf.

8. "Adult Education," National Farmers Union, http://www.nfu.org/education/adult-education.

9. "Iowa Porkettes," Iowa Women's Archives, University of Iowa Libraries, http://sdrc.lib.uiowa.edu/iwa/findingaids/html/IAPorkettes.htm.

10. For a description of the activities of American Agri-Women (2015), see their website at http://www.americanagriwomen.org, which also includes an interesting history of the organization.

11. Although we do not have socioeconomic data on all of the members of PA-WAgN, we do have data from our survey respondents about Pennsylvania farm women's socioeconomic characteristics. Pennsylvania women farmers were more educated than the Pennsylvania population (53.1% with college degrees compared to 26% for the state as a whole) and had higher household incomes (32.9% less than $50,000) than other residents of Pennsylvania (45.7%).

12. PA-WAgN received its first grant from SARE and USDA. See appendix for information on funding sources.

13. A survey of Pennsylvania dairy farmers in 2012 found that only 50% of farmers expressed confidence in their ability to meet their profitability goals in the future. Many anticipated exiting the dairy business due to economic distress or selling their cows and switching to cash cropping by 2017 (Center for Dairy Excellence 2012).

BIBLIOGRAPHY

"2011 Organic Industry Survey." 2011. Organic Trade Association. http://www.ota.com/pics/documents/2011OrganicIndustrySurvey.pdf. Accessed December 9, 2013.

Abraham, M., and B. Purkayastha. 2012. "Making a Difference: Linking Research and Action in Practice, Pedagogy, and Policy for Social Justice: Introduction." *Current Sociology* 60 (2): 123–41.

Acker, Joan. 2006. "Inequality Regimes: Gender, Class, and Race in Organizations." *Gender and Society* 20: 441–64.

Adams, Jane. 1994. *The Transformation of Rural Life: Southern Illinois, 1890–1990*. Chapel Hill: University of North Carolina Press.

Adams, R., and T. W. Kelsey. 2012. "Pennsylvania Dairy Farms and Marcellus Shale, 2007–2010." *Penn State Cooperative Extension*, Marcellus Education Fact Sheet. http://pubs.cas.psu.edu/FreePubs/PDFs/ee0020.pdf.

Agarwal, Bina. 1992. "The Gender and Environment Debate: Lessons from India." *Feminist Studies* 18 (1): 119–58.

Allen, Patricia, and Carolyn Sachs. 2007. "Women in Food Chains: The Gendered Politics of Food." *International Journal of Sociology of Food and Agriculture* 15 (1): 1–23.

Alsgaard, Hannah. 2013. "Rural Inheritance: Gender Disparities in Farm Transmission." *North Dakota Law Review* 88: 347–409.

American Agri-Women. 2015. http://www.americanagriwomen.org. Accessed September 24, 2015.

American Farm Bureau. 2015. "Growing Strong: American Farm Bureau Women's Leadership Program." http://www.fb.org/assets/files/programs/women/FB13_92_WLC_Brochure_f_web.pdf. Accessed September 24, 2015.

Antecol, H., and P. Kuhn. 2000. "Gender as an Impediment to Labor Market Success: Why Do Young Women Report Greater Harm?" *Journal of Labor Economics* 18 (4): 702–28.

Barbercheck, Mary E., Katherine J. Brasier, Nancy Ellen Kiernan, Carolyn Sachs, and Amy Trauger. 2012. "Use of Conservation Practices by Women Farmers in the Northeastern United States." *Renewable Agriculture and Food Systems* 29: 65–82. http://dx.doi.org/10.1017/S1742170512000348.

Barbercheck, Mary E., Katherine J. Brasier, Nancy Ellen Kiernan, Carolyn Sachs, Amy Trauger, and Jill Findeis. 2009. "Meeting the Extension Needs of Women Farmers:

A Perspective from Pennsylvania." *Journal of Extension* 47 (3). http://www.joe.org/joe/2009june/a8.php.

Benbrook, Charles M. 2012. "Impacts of Genetically Engineered Crops on Pesticide Use in the U.S.—the First Sixteen Years." *Environmental Science Europe* 24: 24. doi: 10.1186/2190-4715-24-24.

Beus, Curtis, and Riley Dunlap. 1990. "Conventional vs. Alternative Agriculture: The Paradigmatic Roots of the Debate." *Rural Sociology* 55 (4): 590–616.

Black, A. M. 2007. "What Did That Program Do? Measuring the Outcomes of a Statewide Agricultural Leadership Development Program." *Journal of Extension* 45 (4). www.joe.org/joe/2007august/iw2.php.

Bloor, M., J. Frankland, M. Thomas, and K. Robson. 2001. *Focus Groups in Social Research*. London: Sage Publications.

Bohn, Katrin, and Andre Viljoen. 2011. "The Edible City: Envisioning the Continuous Productive Urban Landscape (CPUL)." *Field Journal* 4 (1): 149–61.

Bokemeier, Janet, and Lorraine Garkovich. 1987. "Assessing the Influence of Farm-women's Self-Identity on Task Allocation and Decision-Making." *Rural Sociology* 52: 13–36.

Bowen, S., S. Elliott, and J. Brenton. 2014. "The Joy of Cooking and Other Cooking Lies." *Contexts* 13 (3): 20–25.

Brandth, Berit. 1995. "Rural Masculinity in Transition: Gender Images in Tractor Advertisements." *Journal of Rural Studies* 11 (2): 123–33.

———. 2002. "Gender Identity in European Family Farming: A Literature Review." *Sociologia Ruralis* 42: 181–200.

———. 2006. "Agricultural Body-Building: Incorporations of Gender, Body and Work." *Journal of Rural Studies* 22:17–27.

Brandth, Berit, and Marit S. Haugen. 2011. "Farm Diversification into Tourism—Implications for Social Identity?" *Journal of Rural Studies* 27: 35–44.

Brasier, K., C. Sachs, N. E. Kiernan, A. Trauger, and M. Barbercheck. 2014. "Capturing the Multiple and Shifting Identities of Farm Women in the Northeastern United States." *Rural Sociology* 79 (3): 283–309. doi: 10.1111/ruso.12040.

Brasier, K., M. Barbercheck, N. E. Kiernan, C. Sachs, A. Schwartzberg, and A. Trauger. 2009. "Extension Educators' Perceptions of the Educational Needs of Women Farmers." *Journal of Extension* 47 (3). http://www.joe.org/joe/2009june/a9.php.

Browne, W. 2001. *The Failure of National Rural Policy Institutions and Interests*. Washington, DC: Georgetown University Press.

Bruni, Attila, Silvia Gherardi, and Barbara Poggio. 2004. "Doing Gender, Doing Entrepreneurship: An Ethnographic Account of Intertwined Practices." *Gender, Work & Organization* 11 (4): 409–29.

Burton, Rob J. F., and Geoff A. Wilson. 2006. "Injecting Social Psychology Theory into Conceptualizations of Agricultural Agency: Towards a Post-Productivist Farmer Self-Identity?" *Journal of Rural Studies* 22: 95–115.

Buttel, F., and J. Goldberger. 2002. "Gender and Agricultural Science: Evidence from Two Surveys of Land—Grant Scientists." *Rural Sociology* 67 (1): 24–43.

Byler, L., Nancy Ellen Kiernan, S. Steel, Patricia Neiner, and D. J. Murphy. 2013. "Beginning Farmers: Will They Face Up to Safety and Health Hazards?" *Journal of Extension* 51 (6). http://www.joe.org/joe/2013december/a10.php.

Carlson, K. F., Gerberich, S. G., Church, T. R., Ryan, A. D., Alexander, B. H., Mongin, S. J., Renier, C. M., Zhang, X., French, L. R. and Masten, A. 2005. Tractor-related injuries: A population-based study of a five-state region in the Midwest. *American Journal of Industrial Medicine* 47:254–264.

Carolan, M. S. 2006. "Social Change and the Adoption and Adaptation of Knowledge Claims: Whose Truth Do You Trust in Regard to Sustainable Agriculture?" *Agriculture and Human Values* 23: 270–85.

Carr, Deborah. 2000. "The Entrepreneurial Alternative." In *Women at Work: Leadership for the Next Century,* edited by D. M. Smith, 208–29. Upper Saddle River: Prentice Hall.

Carson, Rachel. 1962. *Silent Spring*. New York: Houghton Mifflin.

Carter, Angie. 2014. "We Love This Land: Women Farmland Owners and Landscape Change." Paper presented at Rural Women's Studies Association conference, San Marcos, Texas.

Center for Dairy Excellence. 2012. "Pennsylvania Dairy Industry Overview." http://centerfordairyexcellence.org/pennsylvania-dairy-industry-overview/.

Chiappe, Marta B., and Cornelia Butler Flora. 1998. "Gendered Elements of the Alternative Agriculture Paradigm." *Rural Sociology* 63 (3): 372–93.

Colasanti, K. J., C. Matts, and M. W. Hamm. 2012. "Results from the 2009 Michigan Farm to School Survey: Participation Grows from 2004." *Journal of Nutrition Education and Behavior* 44 (4): 343–49. doi: 10.1016/j.jneb.2011.12.003.

Collins, Patricia H. 2004. *Black Sexual Politics: African Americans, Gender and the New Racism*. New York: Routledge.

Comer, M., T. Campbell, K. Edwards, and J. Hillison. 2006. "Cooperative Extension and the 1890 Land-Grant Institutions: The Real Story." *Journal of Extension* 44 (3). http://www.joe.org/joe/2006june/a4.php.

Cone, Cynthia A., and Andrea Myhre. 2000. "Community Supported Agriculture: A Sustainable Alternative to Industrial Agriculture?" *Human Organization* 59 (2): 187–97.

Conglose, J. B. 2000. "The Cooperative Extension Service's Role in Running a Successful County Economic Development Program." *Journal of Extension* 38 (3). http://www.joe.org/joe/2000june/a3.php.

Counihan, Carole M. 2004. *Around the Tuscan Table: Food, Family and Gender in Twentieth Century Florence*. New York: Routledge.

Cowan, Ruth Schwartz. 1983. *More Work for Mother: The Ironies of Household Technology from the Open Hearth to the Microwave*. USA: Basic Books.

Danes, S. M. 1996. *Minnesota Farm Women: 1988–1995; Summary of 1995 Follow-*

Up to 1988 Minnesota Farm Women Survey. Minneapolis: University of Minnesota.

Davidson, Deborah, and William Freudenburg. 1996. "Gender and Environmental Risk Concerns: A Review and Analysis of Available Research." *Environment and Behavior* 28: 302–38.

De Schutter, Olivier. 2013. "The Agrarian Transition and the 'Feminization' of Agriculture." Paper presented at Food Sovereignty: A Critical Dialogue. Yale University, New Haven, CT, September 14–15.

Deere, Carmen D. 2005. "The Feminization of Agriculture? Economic Restructuring in Rural Latin America." OPGP 1. Geneva: UN Research Institute for Social Development. http://www.unrisd.org/publications/opgp1. (2005).

DeLind, Laura B., and Anne E. Ferguson. 1999. "Is This a Women's Movement? The Relationship of Gender to Community-Supported Agriculture in Michigan." *Human Organization* 58 (2): 190–200.

DeMuth, Suzanne. 1993. "Defining Community Supported Agriculture: An Excerpt from *Community Supported Agriculture (CSA): An Annotated Bibliography and Resource Guide*." USDA National Agricultural Library. http://www.nal.usda.gov/afsic/pubs/csa/csadef.shtml. (1993).

Devasahayam, Theresa. 2005. "Power and Pleasure around the Stove: The Construction of Gendered Identity in Middle-Class Hindi South Indian Households in Urban Malaysia." *Women's Studies International Forum* 28 (1): 1–20.

DeVault, Marjorie L. 1991. *Feeding the Family: The Social Organization of Caring as Gendered Work*. Chicago: University of Chicago Press.

Devine, Jenny. 2013. *On Behalf of the Family Farm: Iowa Farm Women's Activism since 1945*. Iowa City: University of Iowa Press.

Dill, Gerald M. 2005. "Glyphosate Resistant Crops: History, Status and Future." *Pest Management Science* 61: 219–24.

Dillman, Don. 2000. *Mail and Internet Surveys: The Tailored Design Method*. New York: John Wiley Co.

Dohoo, I. R., K. Leslie, L. Descôteaux, A. Fredeen, P. Dowling, A. Preston, and W. Shewfelt. 2003a. "A Meta-Analysis Review of the Effects of Recombinant Bovine Somatotropin. 1. Methodology and Effects on Production." *Canadian Journal of Veterinary Research* 67 (4): 241–51.

———. 2003b. "A Meta-Analysis Review of the Effects of Recombinant Bovine Somatotropin. 2. Effects on Animal Health, Reproductive Performance, and Culling." *Canadian Journal of Veterinary Research* 67 (4): 252–64.

Dutcher, James D. 2007. "A Review of Resurgence and Replacement Causing Pest Outbreaks in IPM." In *General Concepts in Integrated Pest and Disease Management*, edited by A. Ciancio and K. J. Mukerjee, 27–43. Dordrecht: Springer.

England, Paula. 2010. "The Gender Revolution: Uneven and Stalled." *Gender & Society* 24: 149–66.

Erickson, Tracy. 2014. "Census of Ag Provides Interesting Statistics." iGrow, June 5.

http://igrow.org/livestock/dairy/census-of-ag-provides-interesting-statistics/#sthash.YFCJajwK.dpuf. Accessed October 7, 2015.

Feenstra, G. 2002. "Creating Space for Sustainable Food Systems: Lessons from the Field." *Agriculture and Human Values* 19: 99–106.

Fenwick, T. 2003. "Innovation: Examining Workplace Learning in New Enterprises." *Journal of Workplace Learning* 15 (3): 123–32.

Fernandez, Raquel. 2007. "Women, Work, and Culture." *Journal of the European Economic Association* 5 (2–3): 305–32.

Ferrell, Anne. 2012. "Doing Masculinity: Gendered Challenges to Replacing Burley Tobacco in Central Kentucky." *Agriculture and Human Values* 29: 137–49.

Fink, Deborah. 1992. *Agrarian Women: Wives and Mothers in Rural Nebraska, 1880–1940*. Chapel Hill: University of North Carolina Press.

Flynn, James, Paul Slovic, and C. K. Mertz. 1994. "Gender, Race, and Perception of Environmental Health Risks." *Risk Analysis* 14: 1101.

Food, Agriculture, Conservation, and Trade Act of 1990 (FACTA). 1990. Public Law 101–624, Title XVI, Subtitle A, Section 1603. Washington, DC: GPO. NAL Call # KF1692.A31 1990.

Freire, Paulo. 1970. *Pedagogy of the Oppressed*. New York: Herder and Herder.

Gassman, A. J., J. L. Petzold-Maxwell, E. H. Clifton, M. W. Dunbar, A. M. Hoffmann, D. A. Ingber, and R. S. Keweshan. 2014. "Field-Evolved Resistance by Western Corn Rootworm to Multiple *Bacillus thuringiensis* Toxins in Transgenic Maize." *Proceedings of the National Academy of Sciences of the USA* 111: 5141–46.

Gauchat, Gordon, Maura Kelly, and Michael Wallace. 2012. "Occupational Gender Segregation, Globalization, and Gender Earnings Inequality in US Metropolitan Areas." *Gender & Society* 26 (5): 718–47.

Geiger, Flavia, Jan Bengtsson, Frank Berendse, Wolfgang W. Weisser, Mark Emmerson, Manuel B. Morales, Piotr Ceryngier, Jaan Liira, Teja Tscharntke, Camilla Winqvist, Sönke Eggers, Riccardo Bommarco, Tomas Pärt, Vincent Bretagnolle, Manuel Plantegenest, Lars W. Clement, Christopher Dennis, Catherine Palmer, Juan J. Oñate, Irene Guerrero, Violetta Hawro, Tsipe Aavik, Carsten Thies, Andreas Flohre, Sebastian Hänke, Christina Fischer, Paul W. Goedhart, and Pablo Inchausti. 2010. "Persistent Negative Effects of Pesticides on Biodiversity and Biological Control Potential on European Farmland." *Basic and Applied Ecology* 11 (2): 97–105. http://dx.doi.org/10.1016/j.baae.2009.12.001.

Gillespie J., R. Nehring, C. Hallahan, C. Sandretto, and L. Tauer. 2010. "Adoption of Recombinant Bovine Somatotropin and Farm Profitability: Does Farm Size Matter?" *AgBioForum* 13 (3): 251–62.

Gold, Mary V. 2007. "Sustainable Agriculture: Definitions and Terms." USDA National Agricultural Library Alternative Farming Systems Information Center. Special Reference Briefs Series no. SRB 99-02. Rev. ed. http://afsic.nal.usda.gov/sustainable-agriculture-definitions-and-terms-1. Accessed November 2014.

Goldberger, Jessica, and Jessica Crowe. 2010. "Gender Inequality within the US Land

Grant University Agricultural Sciences Professorate." *International Journal of Gender, Science, and Technology* 2 (3): 334–60.

Goldsmith, R., I. Feygina, and J. Jost. 2013. In "The Gender Gap in Environmental Attitudes: A System Justification Perspective." *Research Policy: Addressing the Gendered Impacts of Climate Change*, edited by M. Alston and K. Whittenbury. Dordrecht: Springer.

Green Heron Tools. 2015. "DeltaHook Tractor Rapid Hitch." http://www.greenherontools.com/products_tractor_hitch.php. Accessed September 23, 2015.

Gustavson, Jenny, Christel Cederberg, Ulf Sonesson, Robert van Otterdijk, and Alexandre Meybeck. 2011. "Global Food Losses and Food Waste: Extent, Causes and Prevention." Rome: Food and Agriculture Organization. http://www.fao.org/docrep/014/mb060e/mb060e00.pdf. Accessed December 9, 2013.

Guthman, Julie. 2004. *Agrarian Dreams: The Paradox of Organic Farming in California*. Berkeley: University of California Press.

Hambleton, Ruth. 2013. *Annie's Project Celebrates Ten Years*. Ames, Iowa: Iowa State Extension. www.extension.iastate.edu/.../2013/Mar2013ANNIESnews.pdf

Hand, Michael S., and Stephen Martinez. 2010. "Just What Does Local Mean?" *Choices* 25 (1).

Haraway, Donna. 1991. *A Cyborg Manifesto*. New York: Routledge.

Harding, Sandra, ed. 2004. *The Feminist Standpoint Theory Reader*. New York: Routledge.

Hassanein, Neva. 1997. "Networking Knowledge in the Sustainable Agriculture Movement: Some Implications of the Gender Dimension." *Society and Natural Resources: An International Journal* 10 (3): 251–57.

———. 1999. *Changing the Way America Farms: Knowledge and Community in the Sustainable Agriculture Movement*. Lincoln: University of Nebraska Press.

Haugen, Marit S. 1998. "The Gendering of Farming: The Case of Norway." *European Journal of Women's Studies* 5 (2): 133–53.

Hayes, Shannon. 2010. *Radical Homemakers: Reclaiming Domesticity from a Consumer Culture*. Richmondville, NY: Left to Write Press.

Heap, I. 2012. "Global Perspective of Herbicide-Resistant Weeds." *Pest Management Science*. doi: 10.1002/ps.3696.

Heggestad, M. 2015. Home Economics Archive: Research, Tradition, and History (HEARTH). Ithaca, NY: Albert R. Mann Library, Cornell University. http://hearth.library.cornell.edu. Accessed September 23, 2015.

Hesse-Biber, S. N. 2007. "The Practice of Feminist In-Depth Interviewing." In *Feminist Research Practice*, edited by Sharlene Nagy Hesse-Biber and Patricia Lina Leavy, 110–49. Sage Publications.

Hinrichs, C. Clare. 2000. "Embeddedness and Local Food Systems: Notes on Two Types of Direct Agricultural Markets." *Journal of Rural Studies* 16 (3): 295–303.

———. 2003. "The Practice and Politics of Food System Localization." *Journal of Rural Studies* 19: 33–45.

Hinrichs, C. Clare, and Patricia Allen. 2008. "Selective Patronage and Social Justice:

Local Food Consumer Campaigns in Historical Context." *Journal of Agricultural and Environmental Ethics* 21: 329–52.
Holcomb, J. 2008. "Environmentalism and the Internet: Corporate Greenwashers and Environmental Groups." *Contemporary Justice Review: Issues in Criminal, Social, and Restorative Justice* 11 (3): 203–11.
Hoppe, R. A., and J. M. MacDonald. 2013. "Updating the ERS Farm Typology." USDA Economic Research Service. Economic Information Bulletin 110.
Hoppe, R. A., and P. Korb. 2013. "Characteristics of Women Farm Operators and Their Farms." USDA Economic Research Service, Economic Information Bulletin 111.
Hughes, Karen. 2003. "Pushed or Pulled? Women's Entry into Self-Employment and Small Business Ownership." *Gender, Work & Organization* 10 (4): 433–54.
Jarosz, Lucy. 2000. "Understanding Agri-Food Networks as Social Relations." *Agriculture and Human Values* 17: 279–83.
Jellison, Katherine. 1993. *Entitled to Power: Farm Women and Technology, 1913–1963*. Chapel Hill: University of North Carolina Press.
Jones, Robert E., and Riley E. Dunlap. 1992. "The Social Bases of Environmental Concern: Have They Changed over Time?" *Rural Sociology* 57 (1): 28–47.
Jordan, N., R. Becker, J. Gunsolus, S. White, and S. Damme. 2003. "Knowledge Net-Works: An Avenue to Ecological Management of Invasive Weeds." *Weed Science* 51 (2): 271–77.
Key, Nigel, and Michael Roberts. 2007. "Commodity Payments, Farm Business Survival and Farm Size Growth." USDA Economic Research Service. Economic Research Report 51.
Khanal, A.R., Gillespie, J., and J. MacDonald. 2010. "Adoption of Technology, Management Practices, and Production Systems in U.S. Milk Production." *Journal of Dairy Science* 92 (12): 6012–22.
Kiernan, N. E., M. Barbercheck, K. J. Brasier, C. Sachs, and A. R. Terman. 2012. "Women Farmers: Pulling Up Their Own Educational Boot Straps with Extension." *Journal of Extension* 50 (5). www.joe.org/joe/2012october/rb5.php?pdf=1.
Kindon, Sara L., R. Pain, and Mike Kesby, eds. 2007. *Participatory Action Research Approaches and Methods: Connecting People, Participation and Places*. New York: Routledge.
King, Robert P., Michael S. Hand, Gigi DiGiacomo, Kate Clancy, Miguel I. Gomez, Shermain D. Hardesty, Larry Lev, and Edward W. McLaughlin. 2010. "Comparing the Structure, Size, and Performance of Local and Mainstream Food Supply Chains." USDA Economic Research Service. Economic Research Report 99.
Kivirist, L. 2012. "The Secret to HERgonomics." *Hobby Farm Home* (March/April): 87.
Kosteas, V. 2013. "Gender Role Attitudes, Labor Supply, and Human Capital Formation." *Industrial Relations: A Journal of Economy and Society* 52: 915–40.
Kremen, C., and A. Miles. 2012. "Ecosystem Services in Biologically Diversified Versus Conventional Farming Systems: Benefits, Externalities, and Trade-offs." *Ecology and Society* 17 (4): 40. http://www.ecologyandsociety.org/vol17/iss4/art40/.
Lastarria-Cornhiel, Susana. 2006. "Feminization of Agriculture: Trends and Driving

Forces." Working paper no. 41367, background paper for *World Development Report, 2008*, World Bank, Washington, DC.

Lear, L. 1997. *Rachel Carson: Witness for Nature*. New York: Henry Holt and Company.

Leckie, G. 1996. "'They Never Trusted Me to Drive': Farm Girls and the Gender Relations of Agricultural Information Transfer." *Gender, Place, and Culture* 3: 309–25.

Lee, T. Y., Gerberich, S. G., Gibson, R. W., Carr, W. P., Shutske, J. & Renier, C. M. 1996. A population-based study of tractor-related injuries: Regional rural injury study - I (RRIS - I). *Journal of Occupational and Environmental Medicine*, 38 (8):782–793.

Lee-Gosselin, H., and J. Grise. 1990. "Are Women Owner-Managers Challenging Our Definitions of Entrepreneurship? An In-Depth Survey." *Journal of Business Ethics* 9 (4/5): 423–33.

Lewis, J. 1985. "The Birth of the EPA." *EPA Journal* 11 (9).

Lewis, P. 2006. "The Quest for Invisibility: Female Entrepreneurs and the Masculine Norm of Enterprise." *Gender, Work, and Organization* 15 (5): 453–69.

Li, X., M. A. Schuler, and M. R. Berenbaum. 2007. "Molecular Mechanisms of Metabolic Resistance to Synthetic and Natural Xenobiotics." *Annual Review of Entomology* 52: 231–53.

Liepins, Ruth. 1998. "The Gendering of Farming and Agricultural Politics: A Matter of Discourses and Power." *Australian Geographer* 29 (3): 371–88.

Lyson, T. A., and A. Guptill. 2004. "Commodity Agriculture, Civic Agriculture and the Future of U.S. Farming." *Rural Sociology* 69: 370–85. doi: 10.1526/0036011041730464.

MacDonald, J. M. 2008. "The Economic Organization of US Broiler Production." Economic Information Bulletin 38. USDA Economic Research Service. http://ageconsearch.umn.edu/handle/58627.

MacDonald, J. M., E. J. O'Donoghue, W. D. McBride, R. F. Nehring, C. L. Sandretto, and R. Mosheim. 2007. "Profits, Costs, and the Changing Structure of Dairy Farming." USDA Economic Research Service. Economic Research Report 47.

MacDonald, J. M., P. Korb, and R. A. Hoppe. 2013. "Farm Size and the Organization of U.S. Crop Farming." USDA Economic Research Service. Economic Research Report 152. www.ers.usda.gov/publications/err-economic-research-report/err152.aspx.

Martinez, Steve, Michael Hand, Michelle Da Pra, Susan Pollack, Katherine Ralston, Travis Smith, Stephen Vogel, Shellye Clark, Leanne Lohr, Sarah Low, and Constance Newman. 2010. "Local Food Systems: Concepts, Impacts, and Issues." May. USDA Economic Research Service. Economic Research Report 97.

Mathew, A. G., R. Cissell, and S. Liamthong. 2007. "Antibiotic Resistance in Bacteria Associated with Food Animals: A United States Perspective of Livestock Production." *Foodborne Pathogens and Disease* 4 (2): 115–33. doi: 10.1089/fpd.2006.0066.

McBride, W., and N. Key. 2013. "U.S. Hog Production from 1992 to 2009: Technology, Restructuring, and Productivity Growth." USDA Economic Research Service. Economic Research Report 158. http://www.ers.usda.gov/publications/err-economic-research-report/err158.aspx#.Up_bSI1Q375.

McCall, Leslie. 2001. "Sources of Racial Wage Inequality in Metropolitan Labor

Markets: Racial, Ethnic, and Gender Differences." *American Sociological Review* 66: 520–41.

McCright, A. M. 2010. "The Effects of Gender on Climate Change Knowledge and Concern in the American Public." *Population and Environment* 32 (1): 66–87.

McGregor, J., and D. Tweed. 2002. "Profiling a New Generation of Female Small Business Owners in New Zealand: Networking, Mentoring and Growth." *Gender, Work, and Organization* 9 (4): 420–38.

Meares, Alison. 1997. "Making the Transition from Conventional to Sustainable Agriculture: Gender, Social Movement Participation and Quality of Life on the Family Farm." *Rural Sociology* 62 (1): 21–47.

Merchant, Carolyn. 1980. *The Death of Nature: Women, Ecology and the Scientific Revolution*. New York: Harper and Row.

———. 2006. "The Scientific Revolution and *The Death of Nature*." *Isis* 97: 513–33.

Middendorf, G., and L. Busch. 1997. "Inquiry for the Public Good: Democratic Participation in Agricultural Research." *Agriculture and Human Values* 14 (1): 45–57.

Mortensen, D. A., J. F. Egan, B. D Maxwell, M. R. Ryan, and R. G. Smith. 2012. "Navigating a Critical Juncture for Sustainable Weed Management." *BioScience* 62 (1): 75–84.

National Public Radio. 2012. "Women, Hispanic Farmers Say Discrimination Continues in Settlement." http://www.npr.org/templates/story/story.php?storyId=164833428. Accessed September 23, 2015.

Neth, Mary. 1995. *Preserving the Family Farm: Women, Community, and the Foundations of Agribusiness in the Midwest 1900–1940*. Baltimore: Johns Hopkins University Press.

"NOAA Knows . . . Dead Zones: Hypoxia in the Gulf of Mexico." 2009. National Atmospheric and Oceanic Administration. http://www.noaa.gov/factsheets/new%20version/dead_zones.pdf. Accessed December 9, 2013.

O'Donoghue, E., R. Hoppe, D. Banker, and P. Korb. 2009. "Exploring Alternative Farm Definitions: Implications for Agricultural Statistics and Program Eligibility." USDA Economic Research Service. Economic Information Bulletin 49. http://www.ers.usda.gov/publications/eib-economic-information-bulletin/eib49.aspx#.U63MlqhhtCM.

O'Hara, Patricia. 1998. *Partners in Production: Women, Farm, and Family in Ireland*. New York: Berghahn.

Ostrom, M. 2014. "She's My Farmer: Exploring the Role of Gender in the Farmers' Market Movement." Paper Presented at Annual Meeting of Agriculture, Food and Human Values, Burlington, Vermont, June.

Ostrom, M., and D. Jackson-Smith. 2005. "Defining a Purpose: Diverse Farm Constituencies and Publicly Funded Agricultural Research and Extension." *Journal of Sustainable Agriculture* 27 (3): 57–76.

Patton, M. Q. 2002. *Qualitative Research and Evaluation Methods*. 3rd ed. Thousand Oaks: Sage.

Peabody, Mary. 2012. Personal interview.

Penunia, Esther. 2011. *The Role of Farmers' Organizations in Empowering and Promoting the Leadership of Rural Women*. http://www.un.org/womenwatch/daw/csw/csw56/egm/Penunia-EP-12-EGM-RW-Oct-2011.pdf.

Peter, Greg, Michael Bell, Susan Jarganin, and Donna Bauer. 2000. "Coming Back Across the Fence: Masculinity and the Transition to Sustainable Agriculture." *Rural Sociology* 65: 215–33.

Pimentel, D., Sean Williamson, Courtney E. Alexander, Omar Gonzalez-Pagan, Caitlin Kontak, and Steven E. Mulkey. 2008. "Reducing Energy Inputs in the U.S. Food System." *Human Ecology* 36 (4): 459–71.

Pini, B. 2002. "Focus Groups, Feminist Research and Farm Women: Opportunities for Empowerment in Rural Social Research." *Journal of Rural Studies* 18 (3): 339–51.

Pratt, Geraldine. 2010. "Collaboration as Feminist Strategy." *Gender, Place, and Culture: A Journal of Feminist Geography* 17 (1): 43–48.

President's Cancer Panel. 2010. *Reducing Environmental Cancer Risk: What We Can Do Now*. 2008–2009 Annual Report. US Department of Health and Human Services, National Institutes of Health, National Cancer Institute. http://deainfo.nci.nih.gov/advisory/pcp/annualReports/pcp08-09rpt/PCP_Report_08-09_508.pdf. Accessed December 9, 2013.

Pretty, J. 2008. "Agricultural Sustainability: Concepts, Principles and Evidence." *Philosophical Transactions B* 363: 447–65.

Reid, C. J. 2004. "Advancing Women's Social Justice Agendas: A Feminist Action Research Framework." *International Journal of Qualitative Methods* 3 (3).

Rocheleau, Diane. 1996. "Gender and Environment: A Feminist Political Ecology Perspective." In *Feminist Political Ecology: Global Issues and Local Experiences*, edited by Diane Rocheleau, Barbara Thomas-Slayter, and Esther Wangari, 3–23. New York: Routledge.

Rosenfeld, R. 1985. *Farm Women: Work, Farm, and Family in the United States*. Chapel Hill: University of North Carolina Press.

Sachs, Carolyn. 1993. "Rural Women and Environmental Activism." In *Gender and Rurality*, edited by Sarah Whatmore. London: Routledge.

———. 1996. *Gendered Fields: Rural Women, Agriculture, and Environment*. Boulder: Westview Press.

Saugeres, L. 2002. "Of Tractors and Men: Masculinity, Technology and Power in a French Farming Community." *Sociologia Ruralis* 42: 143–59.

Schnepf, R., and J. Richardson. 2011. "Consumers and Food Price Inflation." Congressional Research Service Reports. University of North Texas Digital Library. http://digital.library.unt.edu/ark:/67531/metadc97973/. Accessed September 22, 2015.

Shaw, D., S. Culpepper, M. Owen, A. Price, and R. Wilson. 2012. "Herbicide-Resistant Weeds Threaten Soil Conservation Gains: Finding a Balance for Soil and Farm Sustainability." Issue Paper 49. Ames: CAST (Council for Agricultural, Science, and Technology).

Stets, J. E. 2006. "Identity Theory." In *Contemporary Social Psychological Theories*, edited by P. J. Burke, 88–110. Stanford: Stanford University Press.

Stone, Pamela. 2007a. *Opting Out? Why Women Really Quit Careers and Head Home*. Berkeley: University of California Press.

———. 2007b. "The Rhetoric and Reality of *Opting Out*." *Contexts* 6 (4): 14–19.

Stryker, Sheldon. 1980. *Symbolic Interactionism: A Social Structural Version*. Menlo Park: Benjamin Cummings.

Swackhamer, E., and N. E. Kiernan. 2005. "A Multipurpose Evaluation Strategy for Master Gardener Training Programs." *Journal of Extension* 43 (6). www.joe.org/joe/2005december/.

Tabashnik, B. E., A. J. Gassmann, D. W. Crowder, and Y. Carrière. 2008. "Insect Resistance to *Bt* Crops: Evidence Versus Theory." *Nature Biotechnology* 26: 199–202.

Trauger, A., C. Sachs, M. Barbercheck, K. Brasier, N. E. Kiernan, and A. Schwarzburg. 2010. "The Object of Extension: Agricultural Education and Authentic Farmers in Pennsylvania." *Sociologia Ruralis* 50: 85–103.

Trauger, A., C. Sachs, M. Barbercheck, K. Brasier, N. E. Kiernan, and J. Findeis. 2008. "Agricultural Education: Gender Identity and Knowledge Exchange." *Journal of Rural Studies* 24: 432–39.

Trauger, Amy. 2004. "'Because They Can Do the Work': Women Farmers in Sustainable Agriculture in Pennsylvania, USA." *Gender, Place & Culture* 11: 289–307.

US Bureau of Labor Statistics. 2014. "Women in the Labor Force: A Data Book." Report 1049. http://www.bls.gov/cps/wlf-databook-2013.pdf.

US Census Bureau. 2012. "Survey of Business Owners: 2007 Survey of Business Owners Summaries of Findings." http://www.census.gov/econ/sbo/getsof.html?07women.

US EPA. 2012. "Setting Tolerances for Pesticide Residues in Food." *Pesticides: Topical & Chemical Factsheets*. http://www.epa.gov/oppo0001/factsheets/stprf.htm. Accessed October 22, 2012.

USDA. n.d. "National Organic Program." USDA Agriculture Marketing Service. http://www.ams.usda.gov/AMSv1.0/nop.

USDA. 1997. *Civil Rights at the United States Department of Agriculture: A Report by the Civil Rights Action Team*. Washington, DC: USDA.

USDA. 2009a. *2007 Census of Agriculture: U.S. Summary and State Reports*. Geographic Area Series Publications 1 (51). National Agricultural Statistics Service. http://www.agcensus.usda.gov/Publications/2007/Full_Report/index.asp.

USDA. 2009b. *2007 Census of Agriculture: Women Farmers*. http://www.agcensus.usda.gov/Publications/2007/Online_Highlights/Fact_Sheets/Demographics/women.pdf.

USDA. 2009c. "Small and Home-Based Business: Women in Agriculture." National Agricultural Statistics Service. http://www.csrees.usda.gov/nea/economics/in_focus/small_business_if_women.html. Accessed April 10, 2011.

USDA. 2011a. "Agricultural Chemical Usage: Corn, Upland Cotton, and Potatoes

2010," May 25, 2011. USDA National Agricultural Statistics Service. http://www.nass.usda.gov/Surveys/Guide_to_NASS_Surveys/Chemical_Use/FieldCropChemicalUseFactSheet06.09.11.pdf. Accessed December 9, 2013.

USDA. 2011b. *USDA FY 2011 Budget Summary and Annual Performance Plan.* http://www.obpa.usda.gov/budsum/FY11budsum.pdf.

USDA. 2012a. "Farm Household Well-Being (Historical)." USDA Economic Research Service. http://www.ers.usda.gov/topics/farm-economy/farm-household-well-being/farm-household-income-%28historical%29.aspx#.U3DDEC_R01I.

USDA. 2012b. "Farms, Land in Farms, and Livestock Operations." National Agricultural Statistics Service. Washington, DC: USDA.

USDA. 2012c. "Hispanic and Women Farmers and Ranchers Claims Resolution Process." https://www.farmerclaims.gov/.

USDA. 2012d. "Soybeans and Oil Crops: Background." USDA Economic Research Service. http://www.ers.usda.gov/topics/crops/soybeans-oil-crops/background.aspx#.UqTUjY1Q375. Accessed December 9, 2013.

USDA. 2012e. 2012 Census of Agriculture. NASS. Washington, DC: USDA. http://www.agcensus.usda.gov/Publications/2012/Preliminary_Report/Highlights.pdf. Accessed September 22, 2015.

USDA. 2012f. "USDA Releases Results of the 2011 Certified Organic Production Survey." NASS. Washington, DC: USDA.

USDA. 2012g. "Organic Production." National Agricultural Statistics Service, Economics, Statistics and Market Information. Washington, DC: USDA.

USDA. 2013a. "Adoption of Genetically Engineered Crops in the U.S." USDA Economic Research Service. http://www.ers.usda.gov/data-products/adoption-of-genetically-engineered-crops-in-the-us/recent-trends-in-ge-adoption.aspx#.UqTXr41Q374. Accessed December 9, 2013.

USDA. 2013b. "Corn: Background." USDA Economic Research Service. http://www.ers.usda.gov/topics/crops/corn/background.aspx#.UqTWk41Q374. Accessed December 9, 2013.

USDA. 2013c. "NASS Releases 2012 Chemical Use Data for Soybeans and Wheat." National Agricultural Statistics Service. http://www.nass.usda.gov/Newsroom/Notices/05_15_2013.asp. Accessed December 9, 2013.

USDA. 2013d. "Organic Production." USDA Economic Research Service. http://www.ers.usda.gov/data-products/organic-production.aspx#.Um7DjlN3fTc. Accessed September 22, 2015.

USDA. 2014a. "2012 Census Highlights: Farm Economics." May. http://www.agcensus.usda.gov/Publications/2012/Online_Resources/Highlights/Farm_Economics/#snapshot_sales. Accessed October 12, 2015.

USDA. 2014b. "News Release." USDA Census of Agriculture. Washington, DC. http://www.agcensus.usda.gov/Newsroom/2014/02_20_2014.php. Accessed September 23, 2015.

USDA. 2014c. "2012 Census Highlights: Farm Demographics—U.S. Farmers by

Gender, Age, Race, Ethnicity, and More." May. http://www.agcensus.usda.gov/Publications/2012/Online_Resources/Highlights/Farm_Demographics/#fewer_women. Accessed October 12, 2015.

USDA. 2014d. "2012 Census Highlights: Hog and Pig Farming—A $22.5 Billion Industry, Up 25 Percent since 2007." http://www.agcensus.usda.gov/Publications/2012/Online_Resources/Highlights/Hog_and_Pig_Farming/#farm_characteristics. Accessed October 12, 2015.

USDA. 2014e. "2012 Census of Agriculture: Characteristics of All Farms and Farms with Organic Sales." Washington, DC: USDA.

USDA. 2015a. "Extension." National Institute of Food and Agriculture. Washington, DC: USDA. http://nifa.usda.gov/Extension/. Accessed September 23, 2015.

USDA. 2015b. "Cooperative Extension History." National Institute of Food and Agriculture. Washington, DC: USDA. http://nifa.usda.gov/cooperative-extension-history. Accessed September, 2015.

USDA. 2015c. "US Statistics on Women and Minorities on Farms and in Rural Areas." National Agricultural Library. Washington, DC: USDA. http://www.nal.usda.gov/afsic/pubs/agriwomen.shtml. Accessed September 23, 2015.

USDA. 2015d. "Farm Program Atlas." Economic Research Service. Washington, DC: USDA. http://www.ers.usda.gov/data-products/farm-program-atlas.aspx. Accessed September 23, 2015.

USDA. 2015e. "Settlements and Claims Processes." Washington, DC: USDA. http://www.outreach.usda.gov/settlements.htm. Accessed September, 23, 2015.

USDA. 2015f. "Secretary Vilsack's Efforts to Address Discrimination at USDA." USDA Office of the Assistant Secretary for Civil Rights. http://www.ascr.usda.gov/cr_at_usda.html. Accessed October 13, 2015.

USDA. APHIS. 2003. "Bovine Somatotropin." Washington, DC: USDA Animal and Plant Health Inspection Service.

———. 2007. *Dairy 2007: Part 1, Reference of Dairy Cattle Health and Management Practices in the United States.* Washington, DC: USDA Animal and Plant Health Inspection Service. https://www.aphis.usda.gov/animal_health/nahms/dairy/downloads/dairy07/Dairy07_dr_PartI.pdf Accessed September 22, 2015.

Van Gorp, B., and M. J. van der Goot. 2012. "Sustainable Food and Agriculture: Stakeholder's Frames." *Communication, Culture, and Critique* 5: 127–48.

Whatmore, Sarah. 1991. *Farming Women: Gender, Work, and Family Enterprise*. London: Macmillan.

The White House Council on Women and Girls. n.d. "Women in America: Indicators of Social and Economic Well-Being." http://www.whitehouse.gov/administration/eop/cwg/data-on-women.

White, Joyce. 2004. "Vivianne Holmes' Success with Women in Agriculture." Maine Organic Farmers and Gardeners Association (Spring). http://www.mofga.org/Publications/MaineOrganicFarmerGardener/Spring2004/WAgN/tabid/1347/Default.aspx.

Williams, Christine L., Chandra Muller, and Kristine Kilanski. 2012. "Gendered Organizations in the New Economy." *Gender and Society* 26 (4): 549–73.

Willits, F. K., and N. Jolly. 2002. "Women on Farms: 1980/2001." Paper presented at the Annual Meeting of the Rural Sociological Society, August. Chicago.

Witkin, R., and J. Altschuld. 1995. *Planning and Conducting Needs Assessments*. Thousand Oaks, CA: Sage.

Wittman, Hannah, M. Beckie, and C. Hergesheimer. 2012. "Linking Local Food Systems and the Social Economy? Future Roles for Farmers' Markets in Alberta and British Columbia." *Rural Sociology* 77 (1): 36–61.

Wood, Kathleen. 2013. "Laboring to Learn and Learning to Labor: Experiences of Farm Interns on Sustainable Farms." Masters thesis, Pennsylvania State University.

Xiao, C., and A. M. McCright. 2015. "Gender Differences in Environmental Concern: Revisiting the Institutional Trust Hypothesis in the USA." *Environment and Behavior* 47: 17–37.

Zelezny, L. C., P. P. Chua, and C. Aldrich. 2000. "New Ways of Thinking about Environmentalism: Elaborating on Gender Differences in Environmentalism." *Journal of Social Issues* 56: 443–57. doi: 10.1111/0022-4537.00177.

INDEX

Note: Page numbers in *italics* refer to illustrations.